FUZZY LOGIC

fuzzy logic

DISPATCHES FROM THE INFORMATION REVOLUTION

Matthew Friedman

Véhicule Press

The Canada Council for the Arts since 1957 | Le Conseil des Arts du Canada depuis 1957

Véhicule Press gratefully acknowledges the support of The Canada Council for the Arts for its publishing program.

Cover art and design by J.W. Stewart
Set in Perpetua by Simon Garamond
Printing by AGMV Marquis Imprimeur Inc.

Dépôt légal, Bibliothèque nationale du Québec and the National Library of Canada, second quarter 1997.

CANADIAN CATALOGUING IN PUBLICATION DATA
Friedman, Matthew
Fuzzy logic : dispatches from the information revolution
ISBN 1-55065-088-2
1. Information technology–Social aspects.
2. Information services industry. 3. Information society.
I. Title.
P96.T42F75 1977 303.48 33 C97-900447-0

Published by Véhicule Press, P.O.B. 125, Place du Parc Station, Montreal, Quebec H2W 2M9 http://www.cam.org/~vpress
The *Fuzzy Logic* web site: http://fuzzylogic.total.net

Printed in Canada on alkaline paper.

for Mar

Contents

Acknowledgements

Nothing exists in a vacuum, and this book could never have been written without the great and small contributions of many people. Thanks to Simon Dardick and Vicki Marcok at Véhicule Press in Montreal whose ruthless criticism and confidence in my abilities helped transform my ideas from a manuscript to a real book.

Thanks to Allan Conter for his critical focus and uncompromising clarity; to my friends and colleagues Ron Bel Bruno, Harris Breslow, Robert Seguin, Judith Bernstein and Dreyfuss Fauque, for their insights and stimulating ideas; to David Yates, Rob MacKenzie, Paul Barker and Martin Slofstra for making me a better writer; to Steve Bramson and the staff of TotalNet for Web site hosting and technical assistance.

I owe a particular debt of gratitude to my friend and colleague Mark Shainblum who championed and supported my work, and never hesitated to provide his insight when I really needed it.

Above all, I must thank Marlene Blanshay for her love and support. Her keen wit, perception and penetrating mind inform every word on these pages.

Introduction

> Fuzzy logic is a superset of conventional (Boolean) logic that has been extended to handle the concept of partial truth—truth values between "completely true" and "completely false." It was introduced by Dr. Lotfi Zadeh of UC/Berkeley in the 1960's as a means to model the uncertainty of natural language.
>
> –Mark Kantrowitz, Erik Horstkotte, and Cliff Joslyn, "Answers to Frequently Asked Questions about Fuzzy Logic and Fuzzy Expert Systems." (Internet FAQ)

THE FUTURE IS NOW... at least that's what they tell us.

Future shock... future visions... our popular culture has become obsessed with science fiction fantasies unlike any time since the 1950s. At computer industry trade shows, merchants and manufacturers hawk the future like the latest hot products. Television programs such as Futureworld and Beyond 2000 crowd the airwaves. Newsstand magazines proclaim the dawning of an age out of the back pages of H.G. Wells' *The Shape of Things to Come*—even the March 1997, issue of the public issues magazine *George* featured a "Survival Guide to the Future."

The world of science fiction feels close. During the last two decades, technologies that had existed only in paperback novels and pulp magazines have become commonplace. Work, play and politics, and basic assumptions about information and power have been transformed in a blinding rush of new technologies, and in an effort to assimilate the changes, we cast them in the light of science fiction. In the popular mind they are the technologies of science fiction because only the vocabulary of science fiction adequately describes them. In 1984 author William Gibson wrote in the novel *Neuromancer*:

> Cyberspace. A consensual hallucination experienced daily by billions of legitimate operators, in every nation, by children being taught mathematical concepts... A graphic representation of data abstracted from the banks of every computer in the human system.[1]

Cyberspace is the catchword of our time. Anything related to computers and information technology—that is, anything that carries the cachet of the future—is easily labeled with the prefix 'cyber.' It expresses the same excitement and promise that 'radio'—like the Radio Flyer wagon—and 'atomic'—as in a scene from the 1953 movie The 5000 Fingers of Dr. T in which a little boy creates a device so potent because 'it's atomic!'—did in past generations. 'Cyber-' anything carries the twin validations of science fiction and the marvels of the mythic future in which we live.

In cyberspace itself, the future of science fiction validates the present at every turn. Electronic bulletin board systems, chat channels and Usenet newsgroups—the social meeting places of the techno-savvy—are crowded with characters from science fiction. On any one day, the odds are pretty good that you'll run into an Automatic Jack, Johnny Mnemonic, Hiro Protagonist, Wintermute or Cicada Queen establishing their position on the future's cutting edge by their choice of monikers. For years, I was known in BBS circles as Count Zero, the unlikely hero of Gibson's novel of the same name.

It's ironic, then, that the science fiction future is no more than an extrapolation and a broad satire of the here and now. Gibson himself, who coined cyberspace and is seen by some—much to his displeasure—as the prophet of the future world we apparently inhabit, insists that his novels aren't about the future at all. "I am absolutely not about predicting the future. I just mess with the present," he said in an interview in 1995. *Neuromancer* is Swiftian satire, an absurdist commentary so subtle that his American readers never really got the joke. Through a conjunction of the mundane and the horrifically absurd, the satirist crystalized the then vague trends of the nascent information revolution. "There are those rare moments when you get a sense of what the present is—and those moments for me are what Frederick Jamieson calls the post-modern sublime: a perfect mingling of ecstasy and dread. I can think of no other response to the world we live in."

This is not a book about the future, real or imagined. It is a book about the present—the world we live in. In 1946, in an attempt to illuminate the undefinable source of the writer's motivation, George Orwell noted that "his subject matter will be determined by the age he lives in—at least this is true of tumultuous, revolutionary ages like our own..."[2] Our own time may not be, on the surface, as tumultuous as Orwell's, but we are living in an age of revolution which is social and economic as much as it is technological. We are witnessing the tempestuous birth of an information age, and though we are aware that something is happening, it's still not clear just what it is.

This book is a modest attempt to answer that question, or at least to ask some of the questions that may ultimately lead to some kind of answer. Rapid technological changes have, in a few short years, brought vast information processing power to our desktops. Thay have redefined the way we communicate, and begun to change the commodity basis of the global economy and social interactions. But we are only at the beginning. The direction of the information revolution can only be sketched in terms of potentials and probabilities. The future is as vague, distant and inscrutable as ever, and the impact of the revolution on the present, even the character of the revolution itself, is little more than an indistinct shape projected on a shifting screen of change.

Fuzzy logic is both metaphor and methodology. Comforting binary certainties are unreliable in revolutionary times, as the process of social and technological transformation inevitably subverts the foundation of objective truth. We must contend with the infinite vague categories of possibility, the immeasurable potential paths leading from any point in time. The question "what is happening?" can only be answered with certainty when the revolution is over—and it has only just begun.

Indeed, contingency, subversion and uncertainty have emerged as the only givens of the information revolution. We have come to expect that the only thing that doesn't change is change itself. Everything happens by accident; the moment it is established, conventional wisdom is undermined by some new, unforeseen technological innovation. The use and consequences of our extraordinary new technologies inevitably subvert the purpose for which they were designed.

At the centre of the process, of course, is the Internet, the global matrix of interconnected networks originally built as a United States Defense Department research project that has evolved into the medium and stage of the information revolution. Never intended as an instrument of mass communication or collective intelligence, it has grown into something more, and will surely metamorphize into something yet more unrecognizable. Above all, it illustrates an essential axiom that is true of any technological change, but even more so of this revolution: As Gibson observed, "the street has its own use for things."

And that is the perspective of this book. I'm not interested in the undeniable creative gifts of our great innovators and computer programmers, or in the commercial strategies of Bill Gates. What is important about the information revolution is not the technologies, but how they are used. Only a handful of industrial historians have any interest in just how Johannes Gutenberg's first printing press or the steam looms of 19th-century Lancashire actually worked. These tools, though doubtless fascinating in their design, are most important in their ultimate social and economic effects. It is the same for the astonishing new technologies of our own time.

The information revolution is not being waged by machines, but by people. The front lines are not corporate board rooms, but the millions of offices, storefronts, libraries and desktops that are plugged into the pulse of global change. They are the street, and this book is a report from there. These dispatches from the front lines of the information revolution are an attempt to make sense of what is happening and, hopefully, prepare us for what might come next.

CHAPTER ONE

The Information Revolution

I. The Word Made Flesh

LIKE MOST WRITERS, I fondly remember my first typewriter, a chemical blue Smith Corona electric. It was about 1973, and all that I could think of doing with the machine was to type out a long, detailed and fastidiously-tabbed catalogue of my father's vinyl record collection. I used a now-obsolete technology to create a database of a now-obsolete technology. True, if I had a personal computer and a good database program, the job would have been easier... but that wasn't an option then.

I used that typewriter to write my first short stories, my first poems, and my first attempts at reportage. I remember sitting up, late at night in my first apartment on Duluth street, too poor to afford a real PC and too proud to use the ones at the university, hammering-out immense undergraduate essays on Canadian labour history and "Georg Trakl and the Mythology of Fin-de-Siècle Europe..." And it all would have been easier if I'd had a personal computer and a decent word processor...

There was something satisfying and reassuring in the typewriter—the way each keystroke resulted in a resounding *thwack*, and the way the letters appeared engraved in an ink-ribbon-patterned bas relief on the page. The noise was the one thing that kept me going when I pulled my university all-nighters. You can't fall asleep at the keyboard when you make *that* much racket.

It was an aural landscape of industry. The office was as much an environment of noise as desks and watercoolers. There is a void in modern offices where the noise used to be. Work is being done—probably more efficiently than before—but the industrious sound of a few dozen typewriters crashing away in aleatoric rhythm is missing.

Like many writers of my post-hippie-pre-Generation-X generation, I grew up on a steady diet of Ernest Hemingway stories and movies featuring hard-boiled Hollywood hacks. When I think of a writer, I think of William Holden as Joe Gillis on late-night telecasts of *Sunset Boulevard*, his hands flying over the keyboard of a Remington portable, pulling finished pages with a tearing flourish... and the *sound*, the chorus of little hammers striking paper, violently giving flesh to the word.

II. The Empire of Paper

This great icon of the industrial age passed away in obscurity in 1995, after a long and painful bout with obsolescence. Smith Corona, the last major company to devote most of its energy to the design, manufacture and production of typewriters filed for Chapter 11 bankruptcy protection on July 17, 1995 (companies like IBM, the maker of the once-ubiquitous Selectric, long ago diversified beyond mechanical business machines). Electric and manual typewriters will doubtless be used for decades to come, and after a reorganization, Smith Corona continues to make them... but not in the United States, and not for the American market.

The typewriter, a legacy of the industrial revolution, was a victim of the information revolution, but it, more than any other technology, is what made that revolution possible.

Though the lives of most 19th-century industrial workers were bounded by workshops, mills and mass production, that production became increasingly dependent on information technologies as markets and output grew. However, this growth was accompanied by a corresponding increase in the complexity of industrial management.

More production, bigger markets and the ballooning industrial workforce meant there was an ever greater volume of information. Consequently, a whole cadre of workers emerged whose only role in the production process was managing the information that kept the process running. The increasing bureaucratization of American—and, to a lesser extent European—business was a dominant characteristic of industrial growth in this period. Between 1870 and 1940 the proportion of clerical workers in the overall American workforce rose from less than two percent to ten percent.[1]

Indeed, the steady growth of the clerical caste was a direct response to the objective requirements of the new industrial economy. The railroad, telegraphy, and regular mail service emerged as the communication arteries that enabled companies to expand beyond the narrow constraints of the local market. More significantly, improved communications predicated a radical transformation in the preparation and management of information. Traditional information technologies and clerical practices were inadequate in the accelerating environment of the industrial economy.

The typewriter was the key technology of that transformation, and set the foundation for the creation of an empire of paper. It was the archetype of the office machines that fundamentally altered the workplace. Paper documents of all kinds, memoranda and files distributed functions and responsibilities throughout companies and organizations, and the demand for inexpensive, skilled information workers caused a profound change in the office demographic.

While women were traditionally barred from work in the heavy industry that drove the economy, office work offered a socially acceptable alternative to domestic labour and motherhood. Their smaller hands were better adapted to the cramped typewriter keyboard, and their historical economic disenfranchisement made them a ready pool of inexpensive labour at a time when the demand for office workers was growing exponentially. By the 1890s, the now-familiar female secretary—a scandalous sight only a generation before—had become an essential part of the economy.

Most importantly, the typewriter had a profound impact on the nature of information itself. The industrial economy was built on mass production, mechanization and uniformity of manufacture, but the information super-structure remained essentially a manual endeavour. The bookkeepers and clerks of Dickens' *Hard Times* kept their letters and prepared correspondence with pen and paper—technologies little changed since the monastic scriptoria of the middle ages.

The phenomenal growth that characterized the industrial economy of the late-nineteenth and early-twentieth centuries, despite periodic depressions and recessions, quickly outstripped traditional information management strategies' capacity to adapt. The size and complexity of industry, and the opening of transoceanic markets for industrial goods had made the handwritten ledger and memorandum obsolete. The sheer scale of information management in the industrial economy dictated standardization. In order to bring order to the chaos of paper, heterogeneous accounting systems and file libraries soon gave way to standardized management practices, policies and forms.

The typewriter played a central role in the standardization process. It created uniform, instantly legible text, in much the same way that the assembly lines of the industrial age forged uniform, identical products. Information—in the form of written documents—had become a discrete resource that could be quickly prepared, ordered, stored and quantified.

It didn't take long for standardization to spread to other facets of the information management process. By the mid-1920s, the United States Postal Service had succeeded in establishing standards for envelopes. Manufacturers like Remington and Underwood had re-invested some of the immense profits they had accumulated from the sale of typewriters into the development of increasingly specialized tools, like accounting machines, punch card devices and tabulators—all of which were variations on the typewriter theme.

As the economy grew, so did the use of information processing technologies. In the period between 1922 and 1931, when the adoption of the typewriter and its progeny became almost universal in American business, the United States' gross national product grew to more than $80 billion. The corresponding value of information processing equipment in this period—in terms of the growth of sales, the profits realized by manufacturers and its adoption in the office environment—reflects its importance to economic activity.

Whole industries and professions had emerged that did little more than shuffle papers or provide the tools to do the shuffling more efficiently. Significantly, the document *itself* had become a medium of wealth and power, and corporate empires were built on foundations of stock certificates and bonds. The industrial economy grew on paper as much as on the essential elements of iron, coal and steam, and the exchange of commodities became less a question of shipping physical products from factory to market as a trade of documents.

In-boxes, out-boxes, file boxes and stacks of paper were the pillars of the economy. However, for all of the documents changing hands, it remained a business of industrial production and hard goods. At one level—the highest level—the exchange was between pieces of paper and information, but these were nothing more than abstractions, proxies for concrete industrial products and commodities.

Information (in the form of documents) remained merely an expression of traditional—*real*—commodities and products, and not a commodity itself. A stock certificate had no intrinsic value; it was merely an abstraction of a company's capital and equity, and was traded on the basis of how well the company made and sold concrete products, or how it was expected to in the future. Without the product or capital behind it, the certificate was not, as it was said, worth the paper it was printed on.

The relationship between information and product was dramatically underscored when the American stock markets crashed in 1929. When the exchange of paper outran the supply of commodities and capital it was supposed to represent, the economy collapsed. Divorced from concrete commodities, the information simply had no value.

Information was produced and packaged with the same efficiency as industrial products, but it had not yet itself become the product. As revolutionary as they were, the economic changes wrought by the typewriter did not add up to an information revolution. Such radical transformation would require an extraordinary new tool.

III. Mechanical Reason

Socio-economic transformations start slowly. Someone will wake up with a good idea—a new manufacturing process, a better way of growing crops or an innovation in communications—and give it a try. The first attempts often fail, impeded by inadequate technologies, promotion, or an indifferent market. However, truly revolutionary ideas never quite go away, and when the planets are in alignment and objective conditions—a combination of technological sophistication and social need—are suitable, they achieve critical mass and take off.

The basic principles for the telephone had been well known in engineering circles for a generation before Alexander Graham Bell called his assistant into his workshop in 1877. Charles Bourseuil and Johann Reis had separately experimented with the electrical transmission of sound as early as the mid-1850s. Bell's great contribution was the application of more advanced electrical technology that took the telephone from conception to telecommunications revolution.

Even then the telephone was a severely limited tool, useful only for local communications. Long distance telephone service came relatively slowly, and callers had to wait until 1956—a full century after Bourseuil and Reis' experiments—until they could reach out and touch somebody on the other side of the Atlantic with any reliability.

The steam engine, the *primum mobile* of the industrial revolution, had an even longer gestation. French physicist Denis Papin experimented with a crude engine as early as 1690. After a century of near-misses by some of the Enlightenment's brightest minds, James Watt finally developed an engine that actually worked in 1769. Yet it took sixty years for this miraculous device to find its way into the belly of a commercially viable, practical steam locomotive.

It's hard to say exactly how long ago the idea for a computer flashed into an inventor's mind. Part of the problem is that it is next to impossible to come up with a precise definition of just what a computer is. At a basic level even the most sophisticated supercomputer is nothing more than a numerical calculator—though admittedly an enormously complex one. Computers simply crunch numbers; it's the non-numerical values that programmers and users assign to the numbers that give them any significance.

We have used calculating machines of varying degrees of sophistication for several thousand years. The Chinese abacus, Blaise Pascale's Machine Arithmétique, and Gottfried Liebniz's Stepped Reckoner were increasingly advanced variations on the theme. At the beginning of the nineteenth century, Joseph-Marie Jacquard developed a system of punch cards that allowed industrial weavers to produce identical, pre-set fabric patterns on any number of mechanical looms. Jacquard cloth is a product of the world's first stored memory technology—in effect, the output of a very early and extremely primitive batch program.

For all of their promise, however, none of these magnificent devices could do more than crunch numbers or produce repetitive patterns in cloth. They were all specialized, single-purpose machines, and because information comes in a wide variety of forms—from simple tables of numbers to philosophy dissertations and the elegant sweep of a PostScript font—none really had the potential to manipulate or analyze it in any meaningful way.

The breakthrough came in the mid-1830s when Charles Babbage, an eccentric English mathematician, designed a huge mechanical calculating machine called the Difference Engine. The device, which was never completed, would have been a quantum leap in mechanical computation. As impressive as the Difference Engine would have been, however, Babbage had envisioned an even more remarkable machine that, had it been built, could have brought about an information revolution 140 years before its time.

The Analytical Engine would have been more than a glorified adding machine—though, as with modern computers, arithmetic functions lay at the heart of its design. In her 1842 commentary on the device's design, Babbage's colleague Ada Augusta, Countess Lovelace, identified the feature that lifted Babbage's technology out of the ranks of mechanical calculators and set it firmly in the world of computers: "The engine is capable, under certain circumstances, of feeling about to discover which of two or more possible contingencies has occurred, and of then shaping its future course accordingly."[2]

The Analytical Engine employed a system of "conditional branching," that allows a computer to alter its processes based on the results of previous computations—if it arrives at result w, it initiates operation x; if it arrives at result y, it follows operation z. It is this process that defines modern computers. Rather than simply performing linear calculations, the Analytical Engine could execute its instructions—instructions and data would be fed to it on punch cards borrowed from the Jacquard loom—with hitherto unimagined complexity and versatility without having to be physically reconfigured for each operation.

Only Lady Lovelace fully appreciated what this could mean. One of the most brilliant theorists of her time—the Countess' father, the poet Lord Byron, called her "the Princess of Parallelograms"—she made the conceptual leap from calculating to computing. She alone grasped the idea that, if non-numerical data could be assigned to the numbers, then the Analytical Engine could conceivably compute *anything*:

> ... the Analytical Engine does not occupy common ground with mere "calculating machines." It holds a position wholly its own; and the considerations it suggests are most interesting in their nature. In enabling the mechanism to combine together *general* symbols, in successions of unlimited variety and extent, a uniting link is established between the operations of matter and the abstract mental processes of the *most abstract* branch of mathematical science. A new, a vast, and a powerful language is developed for the future use of analysis, in which to wield its truths so that these may become of more speedy and practical application for the purposes of mankind than the means

> hitherto in our possession have rendered possible. Thus not only the mental and the material, but the theoretical and the practical in the mathematical world, are brought into more intimate and effective connexion with each other. We are not aware of its being on record that anything partaking of the nature of what is so well designated the *Analytical* Engine has been hitherto proposed, or even thought of, as a practical possibility, any more than the idea of a thinking or reasoning machine.[3]

The Analytical Engine would have had a versatility and raw number-crunching power unknown until the mid-1950s,[4] but the world wasn't ready for a "reasoning machine," and Babbage never received the funding necessary to proceed with the project. The fellows of Britain's Royal Society simply couldn't conceive of how such a device might be used, but we can only imagine the frustration Lady Lovelace felt, standing at the very edge of a technological revolution that would not begin for more than a century.

IV. The Universal Machine

It took a world war to make that vision a reality. While no one but Lady Lovelace could conceive of an application for a "reasoning machine" in 1842, by 1942 the military necessity to decrypt coded German messages and documents during the Second World War provided the stimulus for an unprecedented technological innovation.

German messages were encrypted using an ingenious electromechanical device resembling a typewriter called Enigma. A cipher clerk in the field would type a communication and, using a series of rotors, each with 26 possible characters, Enigma would output the coded message. At the receiving end, the coded message was fed back into an Enigma machine, with the rotors arranged in the same sequence, producing legible text. Coded text could be generated by almost any combination of three rotors and up to four spares, meaning that a given character could be one of thirty trillion (3×10^{13}) permutations.

Without the 'key'—the sequence of the rotors—even the best cryptanalyst in the world would be left scratching his head at the gibberish text, and the Germans changed the sequence three times a day. To make matters worse, by 1942 they were using a massive ten-rotor version of Enigma called the

Geheimschreiber—the 'secret writer'—that made their code all but unbreakable. In theory at least, decrypting Enigma was a relatively straightforward question of working out the mathematical formula on which it was based. Given the number of rotors, and enough time, any message could be deciphered. However, in the heat of war, time was something the allies simply didn't have enough of. Strategic planning and the safety of the North Atlantic convoy demanded that the problem be solved as quickly as possible, but Enigma was no simple equation. It was an immense logical problem, far beyond the capabilities of even the most gifted mathematician.

The allied codebreakers pulled out all the stops. The most brilliant mathematicians and logicians on both sides of the Atlantic were assembled to find the key to Enigma. At the top secret British intelligence facility at Bletchley Park outside London, a group of erstwhile university professors, including Cambridge's Alan Mathison Turing, became the central players in one of the most remarkable dramas of the Second World War.

Turing was the kind of man who reveled in abstruse logical problems. In 1936, while a 24-year-old doctoral candidate at Cambridge University, he published a ground-breaking paper on *Computable Numbers*, explored the question of whether it was possible to build a calculating machine to solve all mathematical problems. He concluded that it could not, that there were some logical problems that simply could not be expressed as algorithms.[5]

However, *Computable Numbers* was more than just another theoretical paper, occupying the rarefied stratosphere of abstract thought. It was one of the rare academic papers that had the potential to actually change the world. While proving that a machine capable of solving all problems was impossible, Turing concluded that it was possible to construct a single device capable of solving those problems that were computable. In short, he posited the idea of a machine that could do the work of any calculating machine, that could *be* any machine—a 'universal machine.'

Turing was doubtless familiar with the Analytical Engine. Babbage, after all, had held Cambridge's Lucasian Chair of Mathematics at the time that he was working on the machine's design. However, Turing took the idea a step further. He reasoned that, by breaking complex procedures and data down to their basic steps, any problem that could be solved could be computed by his theoretical machine without physically rebuilding it for each equation.[6]

When he arrived at Bletchley Hall, Turing was in a position to test his theory.

From 1942 onward the British codebreakers focused their efforts on the development of a vast decryption device that could run through every possible combination of the Enigma code until it produced a legible message. It was theoretically possible to have the work done by human 'computers'—mathematicians each assigned to one part of a large-scale calculation—but the sheer scale of the problem meant that such a traditional method would simply take too long. The Colossi, the machines designed by the Bletchley Park team, could do the work of a roomful of 'computers' in a fraction of the time.

The Colossi were single-purpose machines, with no stored memory. They were still a long way from Turing's vision of a universal machine, but they demonstrated that it was possible. While the Analytical Engine would have been a mechanical device, subject to the inherent limitations of moving parts, the Bletchley Park machines were electronic. Numbers were not represented by the teeth of brass gears, but by electronic pulses in 1,800 vacuum tubes, allowing calculations to be performed almost in an instant.

The Colossi couldn't do much more than decipher codes—the task they were designed for. However, most modern computing is little more than code or symbol processing. The Colossi demonstrated that, using the most advanced technology then available, it was indeed possible to perform complex, non-numerical analyses by assigning values to numbers. In that respect, the Colossi were probably the very first electronic computers.

V. ENIAC

On the other side of the Atlantic some of the greatest mathematical minds in the United States were arriving at the same destination, but by a different route. One of the most intractable problems for the U.S. military was calculating the trajectory of artillery shells. The calculations were fairly simple as long as the start- and end-points of a projectile's ballistic curve were stationary, and you didn't take niggling details like wind velocity, friction and drag into account. In the real life and death circumstances of war, these details were far from trivial—they could determine whether or not you destroyed your target before it destroyed you.

As if the problem wasn't complex enough, the age of aerial bombardment, motorized artillery, tanks and aircraft carriers had complicated it further. It's one thing to calculate the theoretical trajectory of a shell fired at a building on a hill from a stationary gun emplacement five miles away; it's something else again to figure out how to shoot down a German jet fighter, traveling west at 600 knots and 20,000 feet, from the deck of an American cruiser traveling 20 knots to the southwest—with a 15 knot westerly wind, and variable air pressure up to the jet's altitude.

The U.S. Army's Ballistic Research Laboratory had employed a variety of calculating devices specially designed by several American universities in the years before and during the war to prepare firing tables that artillerymen could use to aim their guns. However, the scale of the job was enormous, and there was a constant pressure to develop increasingly powerful, ever faster calculating machines. In 1943, the Army commissioned the University of Pennsylvania's Moore School of Electrical Engineering to design an Electronic Numerical Integrator and Calculator—ENIAC—the ultimate ballistics calculating machine.

ENIAC was immense. It occupied 1,800 square feet and weighed 30 tons. Its almost 18,000 vacuum tubes drew 174 kilowats of power, and ran so hot that its designers had to devise a specialized air conditioning system. Yet it performed its task beautifully, and was capable of performing hundreds of multiplications and divisions each second. Unfortunately, it was only ready almost a year after the end of the war.

Like the Colossi, ENIAC was, in essence, a single-purpose calculating machine. For all of its power, it had no stored memory and needed to be reconfigured using patch cables for each new operation. Internally, it boasted a sophisticated logical architecture that, while not entirely mature, had great potential for further development. The machine's chief designers, J. Presper Eckert and John Mauchly understood this better than anyone, and soon left the ENIAC project to exploit its possibilities commercially.

Cambridge University's Electronic Delay Storage Automatic Calculator, or EDSAC, introduced stored memory for programs and data in 1949,[7] and development of ENIAC's successors continued, despite funding problems, at the University of Pennsylvania. Eckert and Mauchly's decision to apply these technologies to a commercially viable product took the computer out of the laboratory, and laid one of the final building blocks for the information revolution.

By today's standards UNIVAC I, the first world's first commercial computer manufactured by the Eckert Mauchly Computer Company, was pretty unimpressive. But it proved that computers had a utility outside of the military or the university lab, and in doing so introduced technologies like magnetic tapes for program and data storage, and input-output devices such as line printers and keyboards that allowed people who didn't happen to be engineers to actually use it.

Most importantly, it demonstrated that there was a market for computers, and by 1953 EMCC had been acquired by Remington Rand, the descendent of the first company to mass produce typewriters. The United States government, the A.C. Nielsen Company and Prudential Insurance all expressed interest in the UNIVAC, and it wasn't long before other companies—notably International Business Machines, a well established manufacturer of typewriters and related office technologies, entered the market with computers of their own.

By the early 1960s computers had become a common tools in business and industry, and pop culture icons in films as diverse as *Desk Set* (1957) and *Alphaville* (1965). The descendants of Colossus and ENIAC were quietly changing the way companies issued invoices and governments audited taxes. The computer promised to change the world, yet change was coming slowly. The information revolution had yet to begin, but the revolutionary vanguard had begun to take its place on the barricades.

VI. The Information Economy

The same wartime research initiatives that gave life to the computer had also transformed the business of information in the United States and Great Britain. It was clear that the war against Germany and Japan had been won by researchers and scientists as much as by soldiers in the field. The scientist, engineer and keeper of information had all become the dominant economic heroes of postwar America, and to a certain extent the British Empire as well.

Information had emerged as a discrete resource in the prewar expansion of the industrial economy. During the war, its production and accumulation—in the form of scientific and military research—had been essential to victory, and this trend accelerated as it became clear that the next world war, should it ever be fought, would be more dependent on technology and innovation than any conflict in history. Between 1940 and 1960 the United States' annual budget for research and development grew from $250 million to $8 billion, an increase of 3200 percent in two decades. By 1972 the budget was $16.7 billion.[8]

The kind of research that this money paid for had been the historic domain of universities and centres for advanced studies, but as the demand for information and innovation increased, government increasingly subcontracted research to the private sector. By 1967, there were no less than 11,355 research firms in the United States. A whole industrial sector had emerged whose sole activity was the production and sale of abstractions—ideas and information.

The nascent information industries enjoyed a symbiotic relationship with the rapidly evolving computer technology. As the volume and complexity of research increased, so did the demand for cheaper, more powerful computers, and the technology itself benefited as private research firms, like American Telegraph and Telephone's Bell labs, produced such innovations as the transistor and the integrated circuit, which in turn made computers both more powerful and affordable for smaller companies.

The cycle of symbiosis began to build towards a critical mass. Computers provided information industries with the ability to accumulate huge volumes of information on one hand, and access to it quickly and economically on the other, and allowed them to provide a wide variety of services. Credit bureaus tracked and sold information on the buying habits and financial reliability of individual consumers. Market research companies like A.C. Nielsen identified economic and political patterns for their growing lists of clients. And at the most basic level of economic activity, a whole cadre of professional consultants emerged to sell their expertise to entrepreneurs and corporate executives often baffled by the increasing complexity of business and technology. In 1979 the French philosopher Jean-François Lyotard observed a profound change in the relationship between information consumers and producers on one hand, and information itself—he used to term 'knowledge'—on the other. Information had acquired the form of value as a commodity, and inevitably, Lyotard concluded, it would be produced and accumulated for the specific purpose of exchange.

Information has been bought and sold in some form for millenia. Whether art and literature, intelligence or scholarly knowledge, there has always been someone willing to exchange it for barter or cash. Indeed, the production and exchange of material products has depended, in large measure, on the parallel exchange of information. The ownership and exchange of land, for example, is expressed in terms of legal title and deed, and all the hog bellies and bushels of grain traded on the global commodities markets exist *at the moment of exchange* as symbolic information. But Lyotard perceived the emergence of a new economic relationship, where information *itself*, not merely as a proxy for material goods or a cultural product, has acquired the status of a commodity.

The traditional symbolic traffic of information has taken on the characteristics of the economy of material products. It is being produced in information industries and traded to accumulate capital. By the 1980s, information had emerged as the dominant force in the invigoration of the American—and by extension the whole Western—industrial economy. Information, and the tools required to process and manipulate it, accounted for 75 percent of the net increase in manufacturing jobs between 1955 and 1979. Spending on high-tech equipment like computers alone accounted for 38 per cent of the United States' economic growth between 1990 and 1994.[9]

Eight of the top 100 American companies listed by *Fortune* magazine in 1970 manufactured computers and related information technologies, but only two, IBM and Sperry Rand, were primarily information technology companies which together had sales totaling $9.2 billion and a combined workforce of more than 350,000 employees. By 1990 ten of the top 100 companies in the United States were in the information technology business, seven of which (IBM, Hewlett Packard, Digital Equipment Corporation, Motorola, Unisys, Texas Instruments and Apple Computer) were principally or solely involved in the manufacturing of computers and their components. They had combined sales of more than $128 billion.

The boom in computer production did not occur in a vacuum. The parallel development of fast and effective data communications technologies meant that not only could information be produced, manipulated and managed—it could also be moved cheaply. Only two telecommunications companies, International Telegraph and Telephone and General Telephone and Electronics, with combined sales of $9.8 billion, were among the top 100 companies in the United States in 1970.

In 1996 there were eight telecommunications companies in the top100, with combined sales of $186 billion, and among the top one-hundred companies in the world, five were telcos, with total sales of $280 billion. The rapid computerization that made this kind of growth possible had the collateral effect of further stimulating the commodification of information. The economies of scale that characterized pre-war industrial production, and the corollary equation that only products whose value exceeds the cost of transporting them to market became moot.

> In the new economy, information in all its forms becomes digital—reduced to bits stored in computers and racing at the speed of light across networks. Using this binary code of computers, information and communications become digital ones and zeros. The new world of possibilities thereby created is as significant as the invention of language itself, the old paradigm on which all the physically based interactions occurred.[10]

The objective conditions for the development of a global information economy—that is, an industrial economy in which the dominant product is information, had been met. At this level exchange is no longer a question of shipping goods to market, but of developing discrete information products and *providing access to them...* at a price, of course. By the mid-1980s, for example, companies offering computerized financial and commercial information to corporate clients were experiencing a business boom.[11]

Information products may be almost anything. The essential characteristic of the information economy is that it is predicated on our ability, using computers, to encode anything, whether it's raw consumer data from a credit bureau or the kind of accumulated expertise of a generation of technical specialists trafficked by consultants. Thus, the information economy encompasses the broadest possible scope of human endeavour. Information products are data, tools, services, experience and entertainment. Anything that can be digitally encoded is information... and anything that is information can be sold.

Consequently, the growth of industries and occupations whose métier is an expression of the new economy has been staggering. While only ten percent of the American workforce were in information related occupations (that is, clerical positions) in 1940, by the mid-1990s almost sixty percent of American workers are involved in the production, accumulation or application of information.[12] Not all of these workers function exclusively as information workers, but information management has become an important part of most occupations. In 1993 the sale of computer software—the most basic information product—accounted for a $2.5 billion chunk of the United States' total exports.

It was the development of a mass market for information products in the 1980s and 1990s that had the greatest impact on the new economy. Confined to the traditional high-tech strongholds of business, research and academia, the development of an information economy would have been little more than an interesting trend, but the emergence of a vast potential market of computerized and technologically-savvy consumers hungry for information of all kinds made it revolutionary.

VII. Revolutionary Vanguard

Of the myths and legends that obscure the early days of personal computers, one of the most evocative is the story of how a teenaged Steve Wozniak, the future co-creator of the Apple computer, once told his father, an engineer at Lockheed, that he would someday own his own computer. The tale is probably apocryphal, but it illustrates the great motivation of one of the most important figures in the development of the PC. At the time that he would have made the assertion, computers were vast, room-sized machines used only by the government, the military and the biggest of the big companies. There were probably no more than 100,000 computers in the whole world at a time when we know Wozniak was sketching his first plans for a computer of his own.[13]

To have conceived at the time that an ordinary individual might someday own a computer of his own, must have appeared to have been madness, or at least evidence of an overactive imagination. But Wozniak wasn't mad, and it was his vision that finally touched off the information revolution. None of that would have happened without the irresistible process of circuit miniaturization that began with the invention of the integrated circuit in 1960. By 1970 engineers at Intel Corporation had succeeded in squeezing all the circuitry of a computer's brain—the central processing unit or CPU—onto a tiny wafer of silicon.

The birth of the microprocessor was one of those little, almost insignificant events that changed the world. A quarter of a century ago, on November 15, 1971, Intel corporation publicly announced the 4004, "a micro-programmable computer on a chip." Though no one used the term at the time, this was the first microprocessor, the direct ancestor of the chip that runs your computer. Intel hyped the 4004 as "a new era of integrated electronics" and the "under $100 computer," but few in the company really understood the impact that this unassuming piece of silicon would ultimately have. Ted Hoff didn't set out to change the world. The Intel engineer who led the 4004 project was simply trying to build a better calculator. In 1969, Busicom, a Japanese electronic calculator manufacturer approached Intel to manufacture the chips for its latest model. "Their original design was too complex, and we found that it wouldn't have met the cost objectives," Hoff said. "There just had to be a better way."

That better way involved combining several chips in one piece of silicon, a simple processor that could be programmed to perform all the functions of a calculator. By using one chip instead of four or five, Hoff was able to reduce the Busicom calculator's power consumption and and cost. "We weren't setting out to start a revolution. I was just an engineer trying to do a good job for a customer." At the time, Intel was primarily a producer of memory chips, but before long the company recognized the potential of its new chip. Intel understood that the 4004 could be used as an embedded controller for cash registers, medical instruments, consumer electronics products, even digital scales at the grocery store.

The only problem was that, as part of the Busicom deal, Intel had given up the rights to the 4004. Without those rights the company would have had to ask Busicom's permission every time it wanted to sell the chip to someone else. The microcomputer revolution might never have amounted to more than a programmable scientific calculator except for one thing—Busicom was in deep financial trouble. Intel, then a tiny Silicon Valley chip maker, convinced the Japanese company to sell back the rights to the 4004 in return for a lower price on the chips. In 1971, Intel boasted that it had crammed 2300 transistors—even more than the building-sized ENIAC could boast—into the 4004. "We thought we were being really aggressive by fitting 2300 transistors onto a single chip," Hoff said. "Today, there are about five million transistors on a Pentium."

In 25 years microprocessors have become more than 2000 times more complex, capable of performing tasks only imagined by science fiction writers.

The size of the transistors has shrunk from ten microns or 0.001 centimetres across on the 4004 to .35 microns, or 0.000035 centimetres on the Pentium. "I can't say we expected that kind of progress anymore than we expected the computer revolution," Hoff said. That revolution took a few years to materialize, but with the success of the first microprocessor, other semiconductor companies like Motorola and Zilog began to develop similar products of their own, and the competition helped spur a feverish pace of innovation. Within six months of the 4004, Intel introduced the 8008, a considerably more complex and powerful microprocessor design and in 1974, had the 8080 chip—one of the first microprocessors suitable for use in a personal computer—ready for market.

The microprocessor changed everything. Its development meant that computers that took up whole air conditioned rooms in the 1960s could be built to sit on a desk. Moreover, so many microprocessor chips could be made at a time that the systems on which they were based could be sold at price that most *consumers* could afford. Someone merely had to put such a machine together. None of the major computer manufacturers of the time, like IBM and Digital Equipment Corporation, chose to pursue this route. The first microprocessor-based *personal* computers were built by freelance enthusiasts, often toiling late at night in their basement and garage workshops. Together with Steve Jobs, Wozniak's high-tech tinkering finally resulted in a real, usable personal computer for the consumer market—the Apple II—in 1977. There had been hobby kits and various failed also-rans for a few years before that, but Apple's sleek little machine was the first personal computer to succeed in the mass market. It was small, affordable, and for its time, relatively powerful.

The widespread use of personal computers was the final necessary condition for the information revolution. Information only began to transform society when the tools became readily available and at least a sizable minority of private citizens—as opposed to institutions and corporations—were equipped to participate in the new economy. It's impossible to say exactly when the number of personal computers in use reached the critical mass necessary to bring information resources into our daily lives, though it's clear that the development of relatively high-performance machines designed specifically for the home market in the late-1980s was a significant factor. By 1995, 34 million U.S. homes were equipped with home systems, representing

35 percent of the total.

Over the course of a century, information had evolved from paper documents produced on typewriters and tabulating machines to become a vast resource comprising the whole of human experience that could be mined, packaged and consumed. However, the burning question is where the information will ultimately lead. Are we destined to remain idle consumers of pre-packaged goods, or are we empowered by the new technology? The information economy, where information is privately held, accumulated and controlled as commodity, and the information society, where it is reserved for the common good, are essentially at odds. In *Information Inequality*, Herbert Schiller correctly argues that the *form* of information and how it is employed "constitute essential and defining features of the social order."

> In the case of information, two dramatically different ways of using it can be imagined. One is to regard information as a social good and a central element in the development and creation of a democratic society. Under this premise, information serves to facilitate democratic decision making, assists citizen participation in government, and contributes to the search for roughly egalitarian measures in the economy at large...[14]

The democratization of information technologies is predicated on commodification. The economies of scale require a mass market for the consumption of information products. However, the thorough commodification of information implicit in the information economy subverts its social form. As a social resource it must be freely accessible. As a commodity it must be scarce, to command a price. An Inuit fisherman would never pay for ice, but the same resource can be as valuable—and expensive—as gold in the Arabian desert. This apparent contradiction is the principal theme of the information revolution. As information has become the dominant element in our economy, the divergent struggles to commodify information or preserve it as a social resource have begun to define our politics and society.

CHAPTER TWO

Everywhere and Nowhere

I. Black Wednesday

DURING A ROUTINE maintenance update and software installation in the early hours of August 7, 1996, something went terribly wrong at America Online. The on-line information service's massive system shut down, and no amount of coaxing and pleading would get it back up. For the better part of the next day AOL's engineers scrambled to find and fix the problem, while its six million subscribers knocked on the door, only to find that no one was home. It was as if a major city had been plunged into a blackout, and though the problem was eventually rectified, its implications would be felt for months afterward.

AOL had begun as one of the many self-contained and hermetically-sealed commercial information networks that vied for consumers' attentions throughout the 1980s and early 1990s. AOL, like H&R Block's Compuserve, General Electric's GEnie, and Prodigy, a service jointly owned by IBM and Sears, offered computer users access to vast databases of information, live news feeds, public discussion conferences and electronic mail within the system—all at exorbitant prices. However, when these commercial services had connections to the outside world at all, either to the Internet or to each other, they were narrow and tenuous. AOL was the digital version of a suburban mall—complete in itself and expensive. All that began to change in the early 1990s. With the computer market, each of the services sweetened their offerings by offering ever more complete access to the global Internet. By the summer of 1996, AOL was the on-ramp of choice to the information highway for six million people.

The biggest news about the AOL crash—or "Black Wednesday," as some users put it—was that it was news at all. For decades, the scattered networks and computer systems that eventually grew into the Internet had, from time to time, collapsed without a murmur in the mainstream press. Among computer scientists, there is almost a perverse faith that every system will inevitably fail, and the process of *trying* to bring down a computer system or network has become an accepted part of system maintenance and disaster management. Yet the AOL crash was different. It was the first time that technical problems at an Internet access provider warranted front-page coverage in the *New York Times*, the United States' newspaper of record, marking the official entry of international computer networking into the mainstream of American business and society.

II. ARPANET

Twenty-seven years earlier while flower children frolicked in the mud at Max Yasgur's farm in Bethel, New York, a team of engineers contracted to the United States Defense Department's Advanced Research Projects Agency struggled to meet a Labor Day deadline to get the embryonic ARPANET up and running. At one level, the project was straightforward enough—an experiment to prove the feasibility of a packet-switching network that would link the computer resources of a handful of geographically separated research institutions. Few of the participants could have known the profound technological, social and economic impact that their experiment would ultimately have. They were building something new, laying the foundation of an information edifice that couldn't have been conceived, not to mention created, only a decade before, and which would soon change the world.

At the time, computers were uncommon, vast, and extraordinarily expensive devices, used only by governments, large corporations and universities. Even the most modest minicomputer sold for tens of thousands of dollars and stood as tall and as wide as a restaurant refrigerator. Access to the valuable machines was strictly limited. Users would often have to type out programs on stacks of index cards, and then pass them on to the select fraternity of computer operators—the priesthood—to have them batch processed. As programs became increasingly complex and demand for computer time rose, it became clear that batch processing methods would have to be replaced by something more flexible. Users needed hands-on access, but to get it they needed a way to let several people use the computer simultaneously. The solution was time-sharing. Because a computer can operate much more quickly that its human operator, researchers reasoned that it could be instructed to jump from one user's programs to another's, and back again, before anyone could tell the difference. Throughout the 1960s they devised ingenious time-sharing systems that maximized computer use by allowing several users at "dumb" terminals to run programs at the same time. Once the exclusive domain of a technological priesthood that had tightly restricted direct access to precious computing resources, time-sharing opened computers up to almost every engineer, professor and graduate student who needed to run a program in real-time, or something like it.

Though time-sharing extended computer use far beyond the glass-house environment to offices and university research labs, it only solved part of the problem. By the early 1960s computers had become an essential part of United States Defence Department-sponsored research. Research facilities around the country, from Stanford University in California to the Massachusetts Institute of Technology, were developing technologies and running experiments—many of which had only tenuous national security implications—on their own systems, but they had no way to communicate and share resources beyond picking up the telephone or meeting at scholarly conferences.

> Researchers were duplicating, and isolating, costly computing resources. Not only were the scientists at each site engaging in more, and more diverse, computer research, but their demands for computer resources were growing faster than [ARPA Information Processing Techniques Office Director] Taylor's budget... And none of the resources or results was easily shared. If the scientists doing graphics in Salt Lake City wanted the programs developed by the people at Lincoln Lab, they had to fly to Boston.[1]

The solution was obvious to the ARPA researchers; create a nationwide research network that would allow computers at facilities around the country to communicate with each other, no matter how far apart they were, or what software they happened to run. The one snag was that no one had ever built such a network before.

ARPANET had another purpose however, one that may not have been explicitly stated in the project specifications, but was nevertheless implicit in how it was deployed. In 1960 RAND Corporation researcher Paul Baran wrote a series of papers detailing a theoretical data communications infrastructure that could survive a nuclear apocalypse. The world was in the frozen depths of the Cold War, and seemed a few short paces from Armageddon at any time. Baran's solution was a completely decentralized network with a large number of redundant connections between each node. Network traffic would be broken up into small, standard-sized 'message blocks' in a technique later known as packet switching,[2] sent through the Net using the best route for each block and assembled when they reached the final destination. "It's like writing a message on a whole lot of postcards," said Internet pioneer Vint Cerf. "You send one postcard by U.S. mail, one by Federal Express, one by bicycle courier, and so on. If one comes back for some reason, you send it by a different route, and once all the parts of the message are collected at the recipient's address, there's a postcard that explains what order you read them in."

When work on the network began in 1968 ARPA researchers adopted Baran's system as a way to keep telecommunications costs low—packet switching doesn't require expensive long-distance telephone lines to be kept open 24 hours per day—and reliability high. Baran had envisaged a system that would automatically re-route messages around nodes that had been destroyed in a nuclear strike, using the multiplicity of redundant connections to keep the line between New York and Los Angeles open by directing traffic through St. Louis if Chicago had been nuked. The same system was ideal for circumventing network congestion and potential network failures, and ARPANET was designed to be robust and reliable, above all—essentially bomb-proof.

At first, ARPANET was a far cry from Baran's vision of a radically decentralized, highly redundant network. The Net began with four nodes at Stanford University, the University of California in Los Angeles and in Santa Barbara, and the University of Utah. By 1971, there were fifteen host sites across the United States, and Raymond Tomlinson had developed the first system for sending electronic mail messages on ARPANET. E-mail was a profoundly significant innovation because while the Net had hitherto seamlessly linked computers, Tomlinson created an instrument that connected people.

From that point on, use of ARPANET grew exponentially. Colleagues on either side of the continent could swap ideas, experimental results, as well as gossip and recipes. The demand for access to this new means of communication was phenomenal.

By 1975 continent-spanning packet switched computer networks had become almost mundane, and there were several self-contained networks, including ARPANET in the United States and Europe in operation or under construction. That year, Cerf and Bob Kahn published the first outline of the Transmission Control Protocol specification that would later evolve into TCP/IP. The idea was to provide a common protocol that would allow all of the various packet switched networks, individually known as internets, to communicate with each other, and in 1983, using TCP/IP, ARPANET became one Net among many in a growing matrix of systems known collectively as the Internet.

From that point on, Internet development continued at a breathless pace. The network infrastructure supported a vast and growing collection of software tools. The most basic ones—like file transfer protocol (FTP) and Telnet—dated from the beginning of ARPANET, but as the Net grew, so did its uses. Electronic mail gave birth to listserv, where groups of users could converse on automated mailing lists, and the interactive public discussion forums of Usenet. In 1990, developers at the University of Minnesota created Gopher, a program that organized any kind of text-based information into a hierarchical menu structure.

Although ARPANET was a defense department project whose day-to-day administration was delegated to the consulting firm of Bolt Beranek Newman in Cambridge, Massachusetts, it was, in effect, run by no one and everyone. True to Baran's vision, it had no central node; every host was as important as the next. Consequently, there was no pre-eminent site that set the standards and established the development agenda. The Net's developers were its users, developing tools as the need arose. If the software worked, it was shared with everyone and adopted by the community as a whole. If it didn't, it was quickly forgotten.

> The tendencies of the ARPANET community ran strongly democratic, with something of an anarchic streak. The ARPANET'S earliest users were constantly generating a steady stream of new ideas, tinkering with old ones, pushing, pulling, or prodding their network to do this or that, spawning an atmosphere of creative chaos. The art of computer programming gave them room for endless riffs, and variations on any theme.[3]

It was organic development, rather than growth by fiat. It had all the character of a community barn raising, with everyone contributing what they could to an infinitely expanding technological structure. Indeed, almost from the beginning, there was a real sense that the Net *belonged* to no one except the users, despite the fact that the Pentagon and its networked research facilities were footing the bill. This nascent on-line culture, based equally on the academic tradition of freedom of information and cooperation, as well as the very distributed topography of the network infrastructure itself, would inform every subsequent development and controversy in the new medium.

III. The Information Highway

Two developments, above all, propelled the Internet from the status of technological curiosity to the essential medium of the information revolution: it opened its gateways to the whole world beyond university computer labs and government-funded research facilities, and it became very easy to use.

With the development of inexpensive personal computers and workstations in the late 1970s and early 1980s, the nature of computing began to change. The ARPANET project had originally been prompted, in part, by the need to provide researchers with access to computing resources at a time when computers were extraordinarily scarce. By the 1980s, however, they were everywhere in universities. The emphasis had changed, and rather than providing access to limited resources, the Net itself had become a vast resource for everyone to share. In 1983 version 4.2 of Berkeley UNIX shipped with Rick Adams' Serial-Line Internet Protocol. SLIP let users connect to Internet hosts through their computers' serial ports, opening up the possibility of cheap access for anyone through telephone lines, from just about anywhere. Dial-up Internet access became increasingly common as Berkeley UNIX quickly became the operating system of choice for university workstations.

The Internet had migrated from computer labs, to faculty offices connected to local area networks, and then finally to homes and remote locations. SLIP brought enormous flexibility to the Net, scaling it down to the personal computers that were just beginning to flood into home offices and living rooms around the world. In 1990, ARPANET—by then just one of many networks woven into the Internet matrix—was officially decommissioned and folded into the regional networks connected to NSFnet, a high-speed packet-switching network launched by the United States government's National Science Foundation in 1986. The Internet was changing. Its value as a military project had faded with the last vestiges of the Cold War, and its expansion beyond the traditional strongholds of research and academia, not to mention the geographical boundaries of the United States, made military sponsorship difficult to justify. NSFnet itself was decommissioned and its backbone handed over to the private sector in 1995.

Nineteen ninety was the year that the Internet began to grow into a mass medium. The on-line community had been an unusual combination of students, researchers, government and military officials since the inception of ARPANET. In 1986, the first Free-Nets, offering some access to Internet e-mail, Usenet and Gopher began popping up in American college towns—usually based in university computer departments. In 1987, SLIP developer Rick Adams founded UUNET, a commercial reseller of Internet services, but it wasn't until The World began the first commercial dial-up service in Boston in 1990, that the Net was set to become the mass medium of the information revolution.

Typically, the Internet explosion happened almost by accident. In 1991 Tim Berners-Lee released an innovative collaboration tool at the CERN high-energy physics laboratory in Switzerland. The World Wide Web was designed as a distributed information publishing and browsing environment, employing hypertext links for navigation. If a physicist was reading a paper and came across a highlighted reference that interested him, he had only to select the link to be transported to the relevant document. In theory at least, Web navigation was simply a question of following a hypertext path of your own choosing. The user, and not the author or network administrator, determined what shape that path would take. In 1993 when the National Center for Supercomputing Applications in Urbana, Illinois, released Mosaic, a client program that reduced Web navigation to pointing and clicking with a mouse in a graphical interface—and included support for images, sound and video besides—the World Wide Web, and consequently the whole Internet, became something that anyone could use. By the end of 1995 more than 30 million people were using the Net.

IV. Communities in Virtual Space

On the Internet the sense of space can be overwhelming. It's a vast expanse of possibilities, an impossibly broad vista stretching to an infinite horizon. The language we use to describe it evokes both place and space. The Internet is an electronic frontier; it's an information highway leading through the far reaches of cyberspace. We speak of sites, storefronts and on-line malls as if the *logical* space of the Net has some material and geographical reality. However, the space that we experience on-line transcends metaphor; it is at the very foundation of interaction in the new medium, William Gibson's consensual hallucination given concrete form in technology.

The Net is the first information medium in which we can interact, asynchronously or in real time, with a community as a whole. The Internet user is both audience and actor. An information consumer is just as easily an information provider. It blurs the line between broadcasting or publishing and conversation. It has volume and place by virtue of the fact that it is a social space, where interactions are mediated in a multitude of dimensions. A one-on-one telephone call, no matter how lively the conversation or how distant the connection, is two-dimensional, a vector joining two distinct points rather than a location. Conversely, the lines of on-line interaction are multidimensional, and mediated by technology. Implicit in the mediation is a nexus or a place. On early electronic bulletin board systems like the WELL, participants found something startlingly new—a geographically diverse yet socially conscious space. As author Howard Rheingold, one of the WELL's earliest participants, observed:

> There's always another mind there. It's like having a corner bar, complete with old buddies and delightful newcomers and new tools waiting to take home and fresh graffiti and letters, except instead of putting on my coat, shutting down the computer, and walking down to the corner, I just invoke my telecom program and there they are. It's a place.[4]

Rheingold attributes the sense of place to the conscious efforts of a computer counterculture that coalesced around the WELL in the 1980s. But the truth is far more basic. Just as three points describe a plane, the Net mediates thousands, even millions of interactive objects, implying—within the medium—an infinite volume of social space.

Given the structure of the medium, it couldn't be any other way. At each step along its evolution, the Internet's a priori purpose as a collaborative research tool has been reinforced. It was a short step from sharing computer resources to sharing information. From there the common exchange of experience and social interaction in a discrete social space was inevitable.

> The power of the Internet is that built into it are tools allowing individuals with similar interests to make contact with one another and share information. Alternatively, once individuals with a common interest make contact through some other medium, they then have a medium through which to collaborate and cultivate a network of others with a similar interest.[5]

The Net is a technology whose ultimate effect has transcended the narrowest interpretation of its initial purpose; and such transcendence is implicit in the very nature of information technologies.

The Internet's collaborative legacy, and even more its basic utility as a technology for the open, non-centralized exchange of the potentially infinite resource of information and human knowledge and experience, has set the tone of on-line interactions, reinforcing its role as a classic 'third place,'[6] a village common of the information age.

This social foundation of the on-line space largely defines the Net's emergent role in the information society. Anthropologist Jon Anderson, who has made an extensive study of cultural groups on-line, has observed that the dominant social dynamic on the Net is familiarity. In the plainest terms, it's an environment where people go to find other people like themselves. It is a comfortable, informal social space, like the local watering hole or social club where co-workers and neighbours congregate after work. When journalist Beth Weise was transferred to a strange city by the Associated Press, she found an experience of *home* on-line. "Alone in a new city," she wrote, "I felt surrounded by a thousand aunts—a thousand aunts with modems."[7]

While the sense of *space* on-line may be a hallucination, an attempt to rationalize interactive abstractions, the Internet's role as a *place* or social space is very concrete. It's a place defined by the quality and nature of the social interaction, rather than geographical location. The Net subverts geography—it's nowhere because location is irrelevant. Conversely, as they say in retail, location is everything, and the Internet is everywhere. Every point is equidistant from every other point on-line, and every off-line geographical location is just as close to the Net. For the increasingly rootless and transient population of North America, the Net is a constant.

Despite a quarter century of attempts at urban renewal, North American cities have been drawn into a seemingly irresistible spiral of decline. In the wake of recessions, corporate downsizing and endless rounds of layoffs, companies have abandoned the city for the green pastures and low rent of suburban industrial parks, taking jobs with them. The suburban migrations that began following the Second World War had an even more dramatic effect. The exodus of the middle class to residential subdivisions promised, in theory at least, everyone a piece of the American (or Canadian) dream in the suburbs. Instead, it impoverished the cities and "had the effect of fragmenting the individual's world."[8] There is no informal public life, no third place in the suburbs. Suburban North America—the seat of the dominant culture on the continent—is a random grouping of unrelated and unconnected private worlds; where *community* exists at all, it exists as an aberration.

Indeed, the phenomenal growth of the Internet in North America—characterized largely by its penetration into middle class homes—is, above all, an explosion into a social vacuum. It is both the primary medium of the information-based economy, and a constant, ubiquitous commons for an information society. Though the information revolution is still in its infancy, the impact of the virtual social space—as much an extension of the information marketplace as the soda fountain was part of the local drug store—has already had a profound impact. It promises a fundamental redefinition of the local community by infinitely extending personal space beyond the tidy boundaries of the backyard fence.

V. The Global City

The Net is a social space without precedent. It is everywhere and nowhere, infinitely malleable and scaleable from thousands of participants in a discussion forum to a handful of intimates in a virtual chat room. The characteristics of any individual's experience of this space—whether it's a 19th century salon, the neighbourhood bar or a brothel—are determined by the content of interaction and each participant's background. The Internet exists as an externalization of human knowledge, scaling the personal experience up to a global scale, and the information revolution to the level of human perception. Indeed, as Marshall McLuhan observed in *Understanding Media*...

> ... the personal and social consequences of any medium—that is, of any extension of ourselves—result from the new scale that is introduced into our affairs by each extension of ourselves or by any new technology.[9]

However, the Net's malleability makes it almost impossible to quantify its scale. It is, at once, a supermarket where the wares of the information economy are bought, sold and transmitted, and a vast resource of knowledge—a common, global memory—to be tapped by individual minds through the medium of technology. It's both mundane and transcendent, expanding the information technologies that, from the written word to television, have extended knowledge beyond insular biological processes and isolated experience. Through the written and printed word, knowledge and information were released from the narrow tribal limits of oral culture. Individual knowledge was emancipated—and isolated—from the experience of community. With electronic media, and particularly with the Net, that process is reversed. The pace and volume of information input put the individual in constant contact with the experience of the rest of the world. All experiences are shared. In short, the Net appears to represent the fullest electronic embodiment of the 'global village,' or, as Rheingold might have it, a global 'groupmind.'

While the idea of the global village seems to resonate loudly on the Internet, the patterns of behaviour are, in fact, more subtle and complex. If it is a village, in the strictest McLuhanist sense, it is one that encompasses many tribes, and if it is a community, it is one with a lot of different boroughs and neighbourhoods. Small, self-contained socio-cultural groupings that exist independently of geographically-based social formations and other similar groups within the on-line community, establishing their own patterns of interaction. They are a kind of virtual creole, internally homogeneous but externally and geographically heterogeneous. "For example, Middle-Easterners who study at North American universities, and contact *each other* on the Internet, aren't at the periphery of the Net, they're typical of it," Anderson said. "You get the same dynamic when a young boy from Illinois goes to MIT and turns into a computer wiz and finds that he blossoms by living on the Internet. The typical feature of the Internet is creolizations."

Interactions *within* Creole micro-communities do adhere to tribal patterns, and each one is a microcosmic global village. The tribal organizations are decentralized and fluid, reflecting the network's logical topology, and prestige within the community tends to be proportional to an individual's participation and the quality and quantity of his information contribution. The micro-communities are abstract spaces defined by what each participant brings on-line, where identity can be continually redefined. A male participant in a virtual chat room devoted to flirting and sexual innuendo once commented "I don't really care whether the person I'm chatting up is *really* a woman. There's no way to know for sure and, in any event, it doesn't really matter. I'm playing with what they bring *on-line*, not with what they are *off-line*." Tribal patterns of organization, reciprocity and prestige, and the direct interactions of oral culture have been rediscovered on-line. At the level of microcosm, the Net is a true global village. At a higher level, it's far more complex.

There is little unity of experience or goals on the Net. Rather, it is an environment encompassing widely differing, competing and often contradictory interests. Civil libertarians battle the forces of traditional authority that long ago abdicated their control of the medium. The corporate leaders of the information revolution compete bit-for-bit with information cottage industries whose economies of scale have been amplified by the technologies of the information revolution. There are fascists, anarchists, families, pornographers, teachers, students, cops and robbers on-line. To suggest that they share common interests, or form a virtual group-mind, or even a dysfunctional family is ludicrous. The millions of Internet users collectively form a community, but the model that best describes it is not the tribal village, but the modern city, with its neighbourhoods, barrios and Latin quarters.

All of these neighbourhoods share a common technological and social infrastructure. Communication and interaction are mediated the same way at every point in the community, regardless of each participant's size and economic weight. It's spurious to speak of anarchy, because there are protocols—in the diplomatic sense—implicit in the system, but it's equally false to suggest that the Net has evolved into a kind of virtual democracy, since the on-line discourse persists without conscious agency. Governments, institutions, telecommunications and high-technology companies may build tools and information highways, but the *content* of the traffic is the organic manifestation of myriad interactions *within* the virtual social space that the infrastructure supports.

The Internet is a conscious and *self*-conscious polity, a social structure mediated by, but independent of the technologies that sustain it. Consequently, it has an existence beyond the forces that control the infrastructure. Governments and corporations may own the wires, but no one owns the Net, and this contradiction—for common good is frequently at cross-purposes with the interests of its virtual landlords—drives the central drama of the information revolution.

VI. A Virtual Marketplace

The millions of individuals and communities on-line are squatters in a space originally designed for institutional research dependent on an infrastructure owned and operated for profit by the leading corporations of the information revolution. The Net polity exists at the sufferance of commerce. According to the New York *Times*, the significance of the AOL crash was not that millions of users had lost access to a virtual social space, nor even so much that a billion-dollar corporation had suffered a technological setback, but that the businesses who used the service to participate in the virtual marketplace had been inconvenienced. The Net provides the information economy with the ideal medium for the transmission of information products, and its users are an ideal market. Indeed, the conventional wisdom holds that electronic commerce—that is, commerce conducted through electronic media—will revolutionize the Internet and, by extension, the world. As information products become the dominant commodities of the global economy, and as a larger proportion of consumers find their way onto the Net, it stands to reason that an increasing volume of spending will occur on-line.

Visa and MasterCard, who together account for almost a billion of the credit cards in worldwide circulation, are betting on an electronic shopping bonanza with an unlikely partnership between them and arch-rivals Netscape Communications and Microsoft Corporation to establish a standard for secure on-line transactions. In order to alleviate consumers' concerns about using their credit cards on the Internet, Visa and MasterCard jointly announced the development of a security standard for credit card-based transactions on the Internet in 1996. Employing advanced encryption technologies, the Secure Electronic Transactions standard shows every sign of becoming the Net's de-facto standard for commercial exchanges. So-called cyberbanks and credit card proxy systems have risen and fallen on-line with the regularity of the seasons, but the entry of the credit card giants into Net commerce seems to promise that the on-line community will be as much a commercial space as a social one.

In fact, electronic commerce has long been a reality at the higher levels of the economy. The trillions of dollars that change hands every day on stock exchange trading floors and in the international commodity and currency markets exist only as bits stored on bank computers, circulated around the world at the speed of light on fibre optic networks. A New York broker can trade on Hong Kong's Hang Seng exchange without leaving Wall Street. Indeed, electronic commerce is the lifeblood of the global economy, and with millions of people signing up with Internet service providers every year, it was only a matter of time before it extended to the consumer trade. Traditional mail-order outfits like Land's End have been tripping over themselves to bring their long-established catalogues to millions of credit-card equipped Net surfers, and the millions who will surely follow in their wake. You still have to wait for the FedEx guy to turn up at your door with your order of 'genuine chinos,' but for anyone who loathes fighting the Saturday afternoon crowds in a department store's bargain basement, on-line commerce is a breath of fresh air.

Commercial exchanges now occur in the on-line ether thousands of times every day, but they remain grounded in the mundane world of off-line banking. Banknotes and pennies don't change hands, and the deal is consummated with a digital handshake, but at some point you still have to pay the bill—and interest—with real currency at a real bank in the real world. However, as on-line commerce evolves, this kind of inelegant electronic payment system can only go so far, and the next step in electronic commerce will require a currency revolution.

In *Being Digital*, Nicholas Negroponte mused on the inevitable arrival of micropayment commerce. He reasoned that, without the imperative for an expensive physical distribution, information producers could charge pennies for their wares—say a newspaper article or a java applet—and make a huge profit on the number of buyers alone. Instead of paying $1.50 for a Saturday newspaper, for example, you could download those few articles that really interest you for a few cents apiece. With a wide enough potential market—

say a few million readers—it could potentially reap a tidy profit through micropayments, but the implicit economies of scale simply don't make any sense with traditional payment systems. After all, cheques and credit cards have service charges for every transaction. Rather, what electronic commerce needs to really succeed is some kind of digital analog to the dollars, sawbucks and quarters jingling in our pockets. The solution isn't credit card security, but electronic cash.

E-cash isn't really much of a stretch beyond the electronic banking services that have become commonplace. The idea is for banks to issue funds for use solely in electronic transactions. If you had $1000 in the bank, you could 'withdraw' $50 to an electronic 'wallet' that would be debited for every transaction. When you had spent all the money in your wallet, you could always return to the bank for more. The promise of such a payment system is that, suitably enough for the Internet—it would operate without intermediaries like credit card companies on a peer-to-peer basis—just like real cash. Digicash BV, an Amsterdam-based firm that pioneered a smart-card based automatic road toll payment system for the Dutch government, and held an experimental pilot project for just such an electronic currency system on the Internet in 1995. Consumers with accounts at participating banks pay for on-line products and services in an independent currency called Ecash, and Digicash-equipped merchants were paid directly by participating banks.

At a fundamental level, all currency is an abstraction of the exchange value of products and labour anyway, so one more level of abstraction shouldn't be a problem... but it is. The Japanese economist Tatsuo Tanaka has raised some of the questions we should have been asking all along. For example, the defining characteristic of electronic commerce—and thus electronic currency—is its fluidity. Unlike real, off-line currency, e-cash knows no borders. Its great value is that any amount of it can be moved halfway around the world in the blink of an eye... but that's just the problem. E-cash is transnational and almost infinitely fluid, thus subverting the national character of off-line currency systems and making it impossible for national central banks to control the circulation and accumulation of huge amounts of untraceable capital. The repercussions for national economic sovereignty could be staggering.

The micro-economic implications of a small proportion of the populations of North America and Europe buying chinos from Land's End aren't even worth worrying about, but it is far more than that. The Net is growing at a phenomenal rate; according to one study, it has doubled in size from 6.6 million hosts in mid-1995 to 12.8 million hosts in mid-1996. Since each host represents more than one user, and perhaps as many as four, the total worldwide population in 1996 must have been substantially more than 35 million users—doubling every year!

Should the number of users and the geographical spread of the Net rise to even five percent of the global population—even with much higher concentrations in the western industries—it could reach an electronic commerce critical mass of staggering implications to monetary stability of which we are but dimly conscious. The growing market for information products and services, and its vast geographical expanse makes some form of e-cash inevitable, and its blessings will be mixed. Electronic currency is as easily laundered and concentrated by individuals as by corporations. Cash, by definition, is next to impossible to trace, and the electronic variety is theoretically fluid enough to be transferred off-shore at the first sign of civil disturbance. Despite that, it also promises to open up the production and markets of the North American and European industrial heartland to the margins of the global economy.

CHAPTER THREE

Global Revolution

I. The Pacific Age

IT WAS BUSINESS AS USUAL for the presidents, prime ministers and other dignitaries whose flag-draped limousines drove through the streets of Manila at the end of November 1996. Representatives of the 18-member Asia-Pacific Economic Cooperation Forum, or APEC, had come to shake hands, make deals, advance their countries' economic agendas and incidentally promote the common economic interests of the nations rimming the Pacific Ocean. Despite some early concerns about security, the summit was conducted with the mundane tenor of international trade.

The 1996 APEC summit was extraordinary in its banality. Before 1993 the annual meeting was a relatively minor event on the global talks schedule, attended by junior cabinet ministers and civil servants. For the great industrial powers—the United States, Canada and Japan are the only G7 members that participate in APEC—the Pacific Rim just didn't seem worth getting too excited about, but with the disintegration of the Soviet bloc and the simultaneous start of the information revolution, everything suddenly changed. Regional economies that had long been subordinate to one of the two superpowers were suddenly liberated to explore regional markets, and the germinal infor-mation economy provided both the tools and the product that allowed them to become players on the international stage. Asia, in particular, experienced a period of astonishing growth. The twin political and economic blocs of the Cold War had given way to a multipolar economy, with Asia accounting for one quarter of the world's total economic output, with its gross national product growing at twice the rate as that of the United States.[1]

The leaders of the old industrial order suddenly realized that the global economy had shifted its focus *beyond* North America and Europe. In the 1990s Canada's most important officials—including the Prime Minister and the leaders of all ten provincial governments—made regular trips to Asia with caps in hand hoping to open new markets for Canadian products. In 1993 U.S. president Bill Clinton elevated the APEC forum from the ranks of diplomatic formality to the same exalted level of the G7 summits. With his presence at APEC, Clinton was making it clear that in the new economy, Singapore, Malaysia and Korea are just as important to the United States as Germany and Great Britain.

The principal focus of the 1996 APEC summit was twofold—the international trade in information products, high technology and telecommunications, and discussions of a proposed Pacific Free Trade Zone to take effect by 2010. Asia has emerged as a powerhouse of the information economy. Companies based in Korea, Japan, Taiwan and Singapore—the 'Asian Tigers'—are becoming the dominant producers of the low-cost personal computers on which the information economy depends, and the semiconductors needed to build them. When a Japanese plant which manufactured silicon resin was destroyed by fire in 1995, the resulting shortage of raw materials sent the price of semiconductor chips around the world through the roof. In the Philippines alone, economic growth rose from zero percent in 1992 to 7 percent in 1996, spurred largely by the development of high-technology industries. American and Japanese high-tech companies like Motorola, Texas Instruments, Toshiba and Acer are bringing billions of dollars into the Filipino economy, and Intel alone began construction of a billion-dollar semiconductor plant in 1996.

Free Trade is the necessary consequence of the phenomenal information technology boom in Asia. The world being forged in the information revolution knows no borders. Globalization is the imperative of a new order founded on the exchange of disembodied data products, the trade of abstractions, and the resulting need to find markets to consume and produce information is driving companies like Intel to expand in areas where labour is cheap, raw materials plentiful and growing markets are close by. Global village experienced as a social phenomenon by individuals who, by accident or design, have erected a new on-line polity on the Internet has an economic corollary in the global information market. It is a mistake to see the economic transformation of the information revolution simply as a consequence of the emergence of enabling technologies. They are as much a product of the information economy as its cause.

Just as the Net is simultaneously everywhere and nowhere, the information revolution has begun to break down the distinctions between margin and centre, hinterland and metropolis and factory and market that have been the dominant characteristic of trade for more than a century. It is also a manifestation of the profound economic shifts that have been occurring throughout the industrialized world in the last three decades. While the traditional industrial powers of the North—the United States, Great Britain, Germany, France, Italy, Canada and Japan—dominate production in the information economy, young industrial powers have emerged, seeking to penetrate the global markets on their own terms.

The emerging economies of the south which have traditionally stood at the margins of the global economy, producing raw materials and consuming finished goods, want a piece of the action. While a global economy has existed, to some extent, since Marco Polo's caravan wended into China in the thirteenth century, the globalization of the information age is characterized by an expanding uniformity of production and consumption. Industrial production demands a concentrated and geographically localized work force and fabrication facilities, and the costs of production tend to favour a global division of labour, with the wealthy countries of the north monopolizing production. The information revolution requires none of heavy industry's economies of scale, and information can be produced and processed anywhere, as long as there is a telephone jack and a power outlet handy. More significant, however, is the global homogeneity of mercantile and economic policy. The categories of first, second and third world have begun to fade along with the distinction between margin and centre, creating an economic climate where the doctrines of economic neo-liberalism—like the free market and transnational concentrations of capital— rather than ideology and local interests, define the direction of regional development. Information is the dominant element in the new global economic order, and the *rest* of the world wants to be a part of it.

II. Margin and Centre

Quiet and unassuming, Crispin Makhoka Sikuku doesn't seem to fit the stereotype of a revolutionary in the developing world. He speaks in measured sentences, not in slogans, and his weapon of choice for socio-economic change is not an assault rifle, but telecommunications and information technology. Indeed, Sikuku is something rare in his native Kenya. A systems administrator for ThornTree Communications in Nairobi, he's in the vanguard of the information revolution. In Kenya, as in most emerging economies, the Internet is still an unknown quantity, but the first tremors of the industrialized world's information revolution have reached these far shores, and people like Sikuku are ready to surf the resulting waves. In the developing world the Net is seen as an essential precondition for economic survival. "What we see is that Western civilization has managed to organize information in such a way that it's a resource," Sikuku said. "The Internet will give us access to this highly organized source of information and let us to participate on the information marketplace. It will allow us to fairly rapidly catch up with the rest of the world."

Despite having one of the fastest growing industrial sectors in Africa, Kenya's economy remains impoverished compared to the developed world. Most manufacturing activity is either related to primary resource extraction or food production. Kenya is, above all, an agrarian economy—the vast majority of its people are subsistence farmers and herders—and its $600 million annual trade deficit bleeds capital to the developed countries of the North. Sikuku sees the information revolution as an opportunity for his country to throw off the yoke of economic colonialism, and move from the margin to the centre. "By participating in the information economy and the Internet, we have a chance of some day achieving self-sufficiency," he said. "It is a slim chance. There are many obstacles to overcome before we can fully participate [in the information economy]. The technology promises much, but it is up to us as a people to see that the promise is fulfilled."

Though the information revolution is coming to the developing world, it's coming slowly, and identifying the need for the technology and exploiting it for economic development are different things. "The technology and concept gets to us in hours," said Aamer Choudhary, a New Jersey-based entrepreneur developing Internet services in his native Pakistan. "Unfortunately the implementation takes a lot more time." There is only one telephone for every 130 people in Pakistan, and Choudhary points out that you can wait up to three years for the national telephone service to install a new line in your home or business. To make matters worse, a single 64k channel—less than one twenty-fifth the bandwidth of the average Internet service provider in North America—serves the entire country. By the summer of 1996, Kenya had been connected to the Net for seven months, but because the vast majority of the population doesn't even have electricity, not to mention telephone lines and computers, that connectivity remains superficial. "We'll have full Internet connectivity some day," Sikuku said, "but not in my children's lifetimes."

Choudhary estimates that less than one person in a hundred in Pakistan actually has access to a computer, and the situation is very much the same in Kenya. In the developing world, computers are expensive and hard to come by. What few systems there are are typically used by governments and non-governmental organizations (NGOs) like the United Nations and the Red Cross. Despite the scarcity of technology, developing countries can still benefit from access to the Net. "If you put an NGO onto the Internet, and it can reach 5000 people with its services, then the fact that their lives might be bettered by it is still something worth going for," Sikuku said.

The Reston, Virginia-based Internet Society agrees. Founded by the Internet's pioneers in 1992, the Society has tried to influence the development of the Net, and encourage its implementation around the world. A key part of the strategy to accomplish that is its annual Workshop on Network Technology for developing countries. The week prior to the society's annual conference in Montreal in June 1996, 256 participants from 100 emerging countries were given a crash course in network implementation and management, with the hope that they will use their new-found expertise to establish Internet services in the developing world. "We're trying to accelerate the flow of know-how so they can become better players in this field," said George Sadowsky, the Internet Society's Vice President (education). "When the whole world has the kind of access we have in the United States and Canada, then our job will be done."

While the Internet is sometimes little more than a digital playground or vast virtual shopping mall for many users in North America and Europe, in countries like Kenya—whose per capita domestic product is less than two percent of that of the United States—it is a vital link to the burgeoning global information economy. The developing world has a lot more to gain from the Internet than the developed world. Reliable telecommunications and overnight couriers, while commonplace in North America and Europe, are almost unknown in the developing world. Communicating with the next big city, not to mention with the rest of the world, is a challenge, but the Net has the potential to change all that.

Perhaps unintentionally, the Internet Society workshop bears more than a passing resemblance to nineteenth century missionary strategy in the developing world. The society's instructors are true believers, motivated by a passionate commitment to bring information enlightenment to the world. They reason that by training a handful of specialists from developing countries, they are planting a seed of technological expertise that will spread throughout their countries. Ironically, the plan appears to work. The society can boast that many of its instructors are former students who want to share their knowledge with less experienced students, and Sikuku is committed to spreading the high-technology gospel in his homeland. "The Internet was something I had heard about, but now I have used it, experienced it, and learned how to get it to work for me," he said. "If I can use what I have learned so Kenyans can see that it's possible to use and move information, then I will have done my small part."

III. Information Imperialism or Digital Liberation?

On a visit to Indonesia in 1995, Internet pioneer Vint Cerf stumbled on an the archetype of the information revolution. "I was in a remote village in the jungle, and the last thing I expected to see was high technology," he recalled. Yet Cerf had stumbled upon a computer, hooked up to a Rube Goldberg power source. "I suspect we're going to see a lot more of that kind of thing as people try to work around technological limitations," he said. Most of the world doesn't have electricity, and only ten percent of the world's population have telephones. According to Cerf, half the world's population has made telephone calls, suggesting that there is a great deal of shared access in developing countries. Television has already had a profound impact on the developing world, and it's likely that computers will follow a similar pattern of penetration. "This is a technology that they want, a technology that they need," Cerf said, "and the developing world is going to do everything it can to be part of the information economy."

The emergence of a vast potential market outside or at the margins of North America hasn't escaped the notice of the information and telecommunications industries. The immense potential represented by a world being transformed by the information revolution is too good to ignore. With their dominant position in telecommunications and information technology, American companies are poised to embark on a high-tech gold rush. There is growth everywhere, but particularly in Asia, where the energetic young economies of Singapore, Malaysia, Indonesia, and Korea, have enthusiastically embraced the Net and the mainly American telecommunication companies that promise to plug them into the information economy.

Since the U.S. government decommissioned NSFnet and privatized the Internet's telecommunications infrastructure—the backbone—in 1995, MCI, along with a handful of other powerful telecommunications corporations, has been one of the principal protagonists in the global expansion of the Net. In 1996 the company began an aggressive expansion into Asia, Europe and Canada. Its merger with British Telecom, joint ventures with Stentor Communications, Canada's telecommunications monopoly, Deutsches Telekom and others, and its Concert InternetPlus project in Asia and Europe essentially puts MCI, and not local Internet pioneers like Choudhary and Sikuku, in control of the Net. In a strange twist of technology, the penetration of the Internet *outside* of the U.S. has compelled MCI and other telecommunications giants to undertake a kind of global upgrade. With all lines on the network leading to nodes in the U.S., the telecommunications companies found that global expansion was necessary to preserve their core American market. "There's a constant drum-beat of demand for capacity, especially in the Pacific, where there's a vast expanse of water to be crossed," Cerf said. "Doing these point-to-point hairy billiard-ball designs is really very unattractive. So when we bring up the backbone, there will be nodes in Japan, Australia and probably in a few other key areas so that we can then hub a lot of the traffic into those concentration points. That will simultaneously reduce everybody's average cost and improve the quality of service." It will also make MCI a lot of money, and that, after all, is the point.

With this in mind, the Internet holds both promise and peril for the developing world. Its interactive nature seems to betoken a future of truly free trade in information, with the weaker economies of the developing world being able to circumvent the economic domination of North America and Europe without *necessarily* having to remain in their thrall. The tools and infra-structure of the information economy are being exported in the name of neo-liberalism and the promised freedom may not be so much an equitable access to the global resource of information, but freedom to consume information *products*.

> The globalization that many find such a promising prospect can be viewed more realistically as the phenomenally successful extension of marketing and consumerism to the world community. A world communication infrastructure, heavily dependent on the new information technologies—satellite, computer, fiberoptic cable—is being put in place. It serves largely the needs of global business, engaged in producing and marketing its outputs worldwide.[2]

The information revolution, like the industrial revolution, may turn out to be yet another extension the developed world's economic hegemony, this time with information. "I would have cause to worry," said Sikuku. "You have ten tons of information flowing toward Kenya and a half a ton flowing out. We can't help but be affected. We may not want all of that information, but democracy demands that we be open. And we have so much to gain that this is a risk that we must take."

But how much of a risk is it? The impetus for the global expansion of the Internet has come as much from the traditional economic centres as from marginal countries seeking economic enfranchisement in the neo-liberal free market. Indeed, the current globalization phase of the Internet and the information economy can be traced to a single point, American vice-president Al Gore's public commitment to the export of his country's National Information Infrastructure project at the International Telecommunications Union conference in Buenos Aires early in 1994. Since then, the information revolution—or at least its mercantile message—has radiated from the American economic heartland like 19th-century Yankee Clippers spreading the gospel of neo-liberal free trade. For all his talk of 'bringing all the communities of the world together,' Gore's grand ambitions for a global information highway are not merely the expression of good intentions and international brotherhood. The *primum mobile* of the expansion are the mercantile interests of the mainly American corporations whose transnational expansion is predicated on the existence of a reliable and efficient global information infrastructure. The so-called global information highway is the missing piece of the puzzle, a prerequisite for the a corporate world order in which the global economy is run by a relatively small number of transnational corporations.

MCI isn't investing billions of dollars in infrastructure upgrades throughout Asia and Latin America to make life easier for Indonesian villagers who want to surf the Net, though it may ultimately have that effect. Corporations spend money to *make* money, and their incentive to expand globally is dictated by necessity to create new markets. If the lives of the potential consumers happen to be improved in the process, that is a happy collateral—though inessential—benefit. From this perspective, it's hard not to view the extension of the tools and media of the information economy as a sort of latter-day imperialism with a smile.

As with the expansion of the industrial economy to the margins in the 19th and early-20th centuries, the benefits to the emerging—and developed—countries may come at the cost of their economic autonomy. Their great dilemma is that, while they must embrace the information revolution to participate in the global economy, it seems to mean accepting the transnationals' technological domination. In effect, they may send out virtual trading fleets, but the ships are owned and operated by foreigners—the same people whose products and services they will buy. The increased transnational flow of data strengthens the world business system's control over local markets and producers. Paradoxically, the nascent global information infrastructure is both the tie that binds marginal economies to the corporate interests of the centre and the medium to share the North's affluence. The challenge for the developing world, and also for the developed economies in the shadow of American corporate power, is to keep the latter open while wriggling out of the restraints of the former.

The central theme of the information revolution is subversion, that 'the street has its own use for things,' and it may be possible to embrace the made-in-the-USA technologies without succumbing to its corporate imperative. One of the great ironies of the Internet Society's developing world workshop is that, though it is largely underwritten by the same American corporations pursuing the manifest destiny of the information highway, it may well prove to be a bulwark against information imperialism. The workshop's ultimate objective is not unlike arming the natives with repeating rifles so they can defend their land against imperialist fortune hunters. While the industrialized North has historically tended to use its technologies for its own benefit while keeping it out of the hands of the local populations, teaching the citizens of developing countries how to use the technologies themselves makes this somewhat less likely in the information age.

That's just what Sikuku has in mind. Though tempted by the fruits of the information revolution, his goal is to help ensure the thorough deployment of the technology in Kenya by Kenyans before someone does it *for* them. "There is an urgency to get this thing done," he said, " but it is something we must do with caution. We do not want to fall impossibly behind, but we are behind already, so we must find our own way." In effect, Sikuku believes that the technology must go native before his country commits itself to the global information economy. "The pressure to accept only what technology they want to give us seems a lot like imperialism. But it's not imperialistic for the United States to say 'come and learn about the Internet.' It's important for us to come, and plant what we have learned back in our own countries." The challenge for the parts of the world only now emerging into the information revolution will be to subvert the transnationals' global expansion by empowering themselves with the tools of the information economy. The revolution must come home.

IV. Northern Exposure

As countries around the world struggle to embrace the information revolution while holding off the forces of information imperialism, one of the most successful models of regional empowerment has emerged right at the edge of the heartland of the new economic order. With 15,000 residents Yellowknife is the largest town in the Canadian North. Most Arctic communities are far smaller and even more remote. Roads are almost unknown in the Arctic, and a trip to the next town usually involves a gruelling and extraordinarily expensive flight in a small, propeller-driven airplane. This is a part of the world where basic phone service—let alone cable TV or net access—is not a given. An uncooperative geography characterized by vast expanses of barren tundra, and rock-hard permafrost, makes laying ordinary telephone cables an exercise in engineering ingenuity, if not downright impossible. Yet the Canadian North is becoming one of the most thoroughly wired places on the planet—a Silicon Tundra.

It all began with the "Connecting the North" symposium in November, 1994. Organized by the Inuit Broadcasting Corporation and the government of the Northwest Territories' Department of Education, Culture and Employment, the conference was a uniquely northern exercise, bringing together government, business and the communities on a scale unparalleled anywhere else in the world. Hundreds of participants from across the vast expanse of the

Canadian North participated in person, and by telephone and live satellite video broadcast. The information revolution has come to the North with all the urgency of an Arctic spring, but, in the wake of Connecting the North, that revolution has taken on a decidedly northern character. The foreword to *Northern Voices on the Information Highway,* the symposium's final report, made it clear that the impetus for the Arctic information revolution would come from the North itself.

> We Northerners have always adapted new technologies to enhance and preserve our unique cultures and traditions. The information highway is a new communication tool that we must embrace and tailor to meet our needs... Northerners must continue to determine their own destiny. By strengthening our ability to communicate amongst ourselves and with the rest of the world we will be better able to participate in the global community.[3]

Arctic Canada is ready to welcome the Internet and everything that comes with it, but only on the North's own terms. *Northern Voices* is its revolutionary manifesto.

Geographic isolation is a fact of life in the Arctic, but takes a great toll on northern communities. Inter-community communications are almost as expensive and difficult as connections with the big cities to the south. A majority of the 100,000 residents of Canada's North—the area north of the 60th parallel—are aboriginal, and most are under the age of eighteen. It's this demographic group that is particularly affected by the isolation, remoteness and emptiness of the Tundra.

Employment opportunities are typically few and far between. Consequently, northern youth are the most under-employed group in Canada. Young people in remote communities often have to choose between their friends, family, and culture or a job in a southern city. Tania Koenig, site administrator of the Aboriginal Youth Network (AYN), a virtual forum for native youth that is funded by the Canadian federal government's Department of Health, points out that many social problems affecting northern communities—including extremely high youth suicide rates and substance abuse—are directly related to isolation and unemployment. "The most common complaint of aboriginal youth in remote communities is that there's nothing to do," she said. "They're bored, and boredom leads to depression." The suicide rate among aboriginal youth is the highest in Canada.

Most of these remote communities are without banks, libraries or hospitals—institutions often taken for granted in even the smallest towns in the United States and southern Canada. "One of the defining aspects of the north is the difficulty of communicating with other communities," said Rici Lake, coordinator of Distance Learning Systems for the government of the Northwest Territories (NWT), and co-founder of NTnet, the Territories' nonprofit Internet hub. "When you have a community of 300 people, what you see is what you get. There's no one in the surrounding tundra."

Since the Connecting the North symposium, the technology envisioned has begun to work a profound transformation on life in the north. One of the NWT government's main goals is to have every citizen connected to the Internet by 1999. In Yellowknife itself, every school is linked to a central server by a microwave network. Ten Internet service providers are staking their claims in the north. Many are operated by entrepreneurs, eager to cash in on what's probably the biggest business opportunity in the north since gold was discovered in the Yukon. In Iqaluit—the largest city in the eastern Arctic, with 3,000 inhabitants—300 subscribers, or 10 percent of the population, have signed up with the local ISP in its first six months of operation.

"It's not an exaggeration to say that the Internet has far greater value to residents of a northern community than in Toronto or Montreal," said Tim Stupich, policy manager at Aboriginal Business Canada, a project of the Federal Department of Industry. "The technology has the potential to completely break down that geographical isolation." The net promises to give northerners access to services in the next town or anywhere in the world. It can bring banks and libraries to people who have never been able to use those services before. "This isn't just a convenience—in a very real sense, it's a necessity," Lake said. Unlike traditional information media such as television, which simply bombard Arctic communities with news of the world outside, the Internet is interactive. Thus, it may be the ideal medium to bring the culture, creativity, and ideas of the north to the world.

Precious metals and Inuit art, which account for most of the northern export market, are practically the only commodities with dollar values that exceed the cost of transporting them to markets in the south. Consequently, the Internet exhibits cost-efficiency as an information conduit, promising greater potential to revitalize the Arctic economy than any other technology. "One of the most significant benefits of the Internet is that we are becoming an information exporter," Lake said. "It lets us participate in the global market of information."

The information economy has the potential to be a two-way street. "What we're doing in bringing information to Nunavik, and information from Nunavik to the rest of the world," said Kathy Peloquin, coordinator of the Internet project in the part of northern Quebec known as Nunavik. "The small enterprises that are successful, are only successful on a shoestring. The reality is that the markets just aren't available to them." Peloquin believes that the Internet is the perfect tool to bring the region's expertise and creativity to the new global market, stimulating the growth of northern knowledge industries. "The content is there, but there has been no way to market it," she said. "You may have something very valuable to contribute—and that people will pay for—but Nunavik's talent has always arrived in the south second-hand, through intermediaries. It has never been marketed directly from the people."

This economic potential has created great excitement in the communities. "We've come a long way in a very short time," said Willy Katainak, mayor of the Nunavik community of Salluit. "Thirty years ago, we communicated by letters, carried by dog team, and now we're getting on the Internet. You can't overstate how important this change will be, especially when you consider that every business in North America is on the Internet—and this gives us access to them."

The creation of such an economy gives new hope to a previously disenfranchised pool of unemployed aboriginal youth. Young people who might otherwise have faced the certain prospect of lifelong unemployment can tap global markets for their creative talents from within their communities. There's no reason to leave home if they can do everything over the Internet. Moreover, native youth who have been drawn south by jobs no longer need be cut off from the social and cultural life of their communities.

An enhanced sense of cultural pride is one of the more abstract bounties of net access in the NWT. The Leo Ussak Elementary School in Rankin Inlet, a community of about 2,000 inhabitants on the northwest rim of Hudson Bay, is only one of a growing number of northern schools putting itself on-line. Computer teacher and systems administrator Bill Belsey is convinced that, by showing their faces to the world and to their neighbors in other remote communities, his students are learning a valuable lesson that they won't get in the classroom. They are reinforcing their sense of pride in their culture and language by holding it up for the world to see on the Net. A series of Internet cafés organized by Belsey significantly stepped up demand for net access. This demand spurred creation of Rankin Inlet's very own Community Access Centre (CAC), where any resident of the town can come to surf the net. Residents have enthusiastically supported the project, realizing quickly that the technology would allow them to become global information producers and not merely consumers.

The Rankin Inlet CAC is the first of such projects, sponsored by Canada's Department of Industry, that are springing up all over the north. More than a dozen are planned for the NWT in the near future, and Taqramiut Nippingat Inc. (TNI), the Inuit-owned public broadcaster in the region of northern Quebec called Nunavik, has established centres in three remote communities, with more to follow. Communities receive a maximum of $30,000 in funding for the project from the federal government, provided that an equal amount is raised locally. That posed no problem to TNI; it has raised three times as much local funding for each CAC.

TNI's Nunavik Net project—launched officially in the fall of 1996—is almost a textbook example of how the information revolution can be directed from *within* a community located at the erstwhile margins of the global economy. Though the project has received some funding from the Canadian government, it is paid for mainly by the community itself and local business interests like the Unaak Shrimp Fishery.

By delegating management responsibilities to young people from the communities themselves the project has the added advantage of empowering local residents with the skills of the information age. The project is a unique combination of public and private funding, cutting edge technology and hard work. There was a lot of improvisation, but the whole project was built from the ground up. The planning and implementation of the project was directed by the communities themselves in consultation with TNI's technicians and engineers. In something of a departure from the way telecommunications technologies have historically been applied to the North, TNI chose to rely on proven, relatively inexpensive high-tech. "In the past, companies approached the region with very expensive, complex equipment," Peloquin said. "We're trying to keep the project as simple—and as reliable—as possible."

Arctic residents simply don't have access to the highly developed telecommunications infrastructure that southerners take for granted. Telephone service is a recent arrival in the North, and rock-hard permafrost and the vast distances between communities have precluded the development of a comprehensive, cable-based network. Peloquin turned to the Boston-based Systems Engineering Society for a solution. SES is arguably the leader in high-speed networking in the North. In the winter and spring of 1995, the company established an experimental network employing air-LAN and satellites in Yukon and the Northwest Territories. When Peloquin approached them with the Nunavik project, SES leapt at the chance to put theory into practice.

"What we have to work with up there is the fact that they just don't have the South's infrastructure," said SES engineer Peter Hinckley. "But that gives us the opportunity to do something that you really can't do in the South. We can provide community-wide ethernet, so we're not just setting Internet access, but the kind of WAN that you'd find in a corporate structure." The two larger communities already have well established cable television networks, owned and maintained by TNI. For Hinckley, that provided an obvious infrastructure for the community WANs. "They already had some cable going into every building," he said. "It didn't make sense to go to the expense of laying new lines when we could use what was already there." Connecting the community networks to the south, however, was another question. All telecommunications are routed through the already overburdened Anik satellites. With no more than 56 kb of bandwidth to work with, the project had to be very creative. "It becomes a challenge to get the most value from the bandwidth. So we're using a combination of techniques, including a compression system that effectively gives us four times the bandwidth, and a store-forward system for large file transfers. We're looking at using a direct PC connection from Telesat for things like the Usenet news feed. That way, we don't have to tie up the 56k for large transfers."

The excitement in the three communities that have been connected so far is overwhelming. TNI has set up community access centres with partial funding from the federal Department of Industry, that have already drawn residents' interest. Most of the funding has come from local sources, with the advantage that the Inuit people have had the dominant hand in directing the project.

The North is betting that its latent—and not so latent—talent and vision need only the right technology to erupt into a creative renaissance. It's clear that the people of Nunavik are eager to jump on-line, but several obstacles remain. Despite the fact that Katainak's children have raised their demands for a home computer to a fever pitch, computers remain a rare commodity in the region. Most Northerners will continue to use the Net from community access centres for some time. This and other shortcomings may prove to be advantageous to the North's technological renaissance. "The centres give people an opportunity to gather together and learn," Peloquin said. "In Puvirnituq we found that it's not just a place to use computers, but a place to develop new ideas. By bringing people together in the centres as well as on the Internet, we can hope to have a fertile creative experience that is unknown in the south."

As serious are the technological barriers created by the unforgiving climate. It's hard to lay cable in the frozen earth, so most inter-community telecommunications rely on microwave transmission and satellite uplinks. "We don't have a southern [telecommunications] infrastructure in the Northwest Territories," Lake said. "In Quebec, the same telecommunications company that serves Salluit also serves Montreal, so consumers in the big city can help subsidize service in the north. However, that's not possible for NWT." On the other hand, the North isn't constrained by established models of connectivity. Northerners aren't tied to telephonic communications, so if the political will exists—and Lake is sure it does—the lack of a land-based infrastructure can be transformed into the north's advantage. "What we're looking at in the Northwest Territory is a strategy that will create inter-community links with a fractional T1 [1.544 mbps], using satellite communications. We've established a baseline of one-half T1 as the projected minimal bandwidth for each community."

The Internet will revolutionize life in the North, offering residents easy access to information resources and services that are either unavailable or outrageously expensive north of the 60th parallel. That much is certain, and efforts to advertise the region as the ideal eco-tourism destination, and to promote the work of local Inuit artists on-line are already well under way.

"It's probably a little premature to start talking about a Silicon Tundra," according to TNI's Tom Axtell, "but there is no place in the world better suited or inclined to use the Internet. Though we're just at the beginning, it's already obvious that this technology is about to utterly revolutionize life in the North—for the better."

It's also true that the technology itself is undergoing a revolution. Where, in the south, the infrastructure is almost entirely in the hands of transnational corporations, the Arctic experience has been almost completely different. The pattern of control and ownership of the telecommunications infrastructure has had to be transformed to suit the North's unique social and geographic conditions. The development of Canada's telecommunications infrastructure has historically been driven by market imperatives. Consequently, it has tended to focus mainly in and between big market centres like Toronto and Montreal. The North has been left out, and residents of remote communities—who *need* access to telecommunications more than anyone else—have generally been unable to take advantage of new services and technologies. Connecting the North concluded that something had to change. The effort to wire the North would not have succeeded if market considerations had not been abandoned, and the extent to which it *has* met with success is a confirmation that an information infrastructure can be established on a non-corporate development model. Lake attributes this to the North's unique cooperative culture.

In an environment where survival itself is often dependent on mutual assistance, a tradition of pooled resources and community action has somehow managed to prevail over the competitive nature of the rest of North America. Even the Northwest Territories' legislature is a non-partisan body, with compromise rather than party politics as its purpose. As Tom Axtell puts it: "There is a Northern way of doing things, and that involves motivating the community *for* the community. As important as the technology itself is, the real story of the Internet in the North is the *way* it has been deployed. This something that the users—individuals as well as companies—own and operate themselves. It's a unique economic model, but it suits our needs. The irony is that, if we weren't so remote, the technology would have been implemented *for us* rather than by us. We are an economic afterthought for most of the world, and that has allowed us to show that there is a different way of doing things."

V. A World Without Walls

For most of the world outside of the United States the economic and social perils of the global information revolution seem, at worst, like the distant thunder of a storm that may, or may not, come their way. Their preoccupations are with language and identity rather than international capital, with cultural sovereignty rather than economic autonomy. As national leaders around the world watch the tide of the information revolution approach at an irresistible pace, the one question they all seem to share is "will our identity and values be drowned in the flood?"

French President Jacques Chirac expressed that sentiment at a summit of French-speaking countries in Benin at the end of 1995. While ready to embrace the immense promise of global networking, he nevertheless cautioned that it could be a vehicle for English, and specifically American, cultural imperialism. The Earth used to be such a tidy place. It was a big planet, with more than enough room for each country to proudly maintain its distinctive identity behind the relative security of its borders. But the Internet has no borders, and countries once separated by political boundaries or thousands of miles of ocean have begun to worry whether the on-line global village is just a synonym for cultural homogenization.

These fears aren't unreasonable in the context of a half-century of American domination of the mass media and cultural industries. The free flow of information is the very foundation of the global information economy and the historical underpinning of cultural domination. In fact, the doctrine has ensured the primacy of American cultural industries. No country's film or media industries can compete with the financial power and marketing budgets of Hollywood and American network television. Where local cultural industries exist at all, they are the poor cousins of their American counterparts, largely denied access to markets in the United States while subtitled Hollywood blockbusters dominate their movie and television screens. Given this historical context—and although more than half of Internet host sites are now located outside of the United States—the Net is often regarded with unease. As it promises to erase geographical distance and bring the whole world into a common marketplace of ideas, it appears to threaten a deluge of foreign, particularly American, culture, values, language and ideas. Judged from the perspective of established information media like television and cinema, the threat to local cultures is dire indeed.

However, the media of the information revolution—particularly the Net—are fundamentally different. The *content* of the Net is created, and not simply consumed, by its users; its is, in that respect, interactive. This added dimension dramatically alters the equation. Though it is conceivable that interactive, global media *may* be dominated by foreign interests—just as an unscrupulous real estate developer could conceivably buy up all of the land in a community's downtown core—such domination is far from certain. In theory the distributed model of Internet communication makes each point on the network as important as any other, making it far more resistant to the kind of control characteristic of more centralized information distribution systems like television.

It is just that decentralization that most worries the world's traditional cultural guardians. As apprehensive as they may be about foreign cultural domination, they are even more concerned about maintaining their own control—or domination—over local cultures. Though Chirac may publicly bemoan the colonizing influence of American Net content, the subtext of his remarks is the abject terror that the institutional arbiters of French culture like l'Académie Française and the French government may become just one voice among many, eroding their control over the language and values of the people. "The Internet doesn't homogenize at all," said anthropologist Jon Anderson, one of a growing number of scholars who specialize in Internet culture. "Homogenizing implies that everyone becomes more the same. What you initially get on something like the Internet are a lot of people who look the same because they are, in a sense, blossoming out there on the Net. What you get is a distinctive Internet culture, and the distinctive Internet culture is international, polyglot... it's cosmopolitan; it's not bound by the rules of existing authority. And from the point of view of Jacques Chirac or an Académie Frenchman, it's bad French."

According to Anderson the growing tension between the guardians of regional cultures and the vast, borderless expanse of cyberspace is inevitable. The Net represents 'the other,' and for governments trying to reconcile the perceived gains of the information revolution with the perceived threat to cultural sovereignty, the problem is formidable. "This kind of instant access to communications and information that is becoming the sine qua non of the industrial world. This really is an information age, but for each country's cultural gatekeepers, the question is how do they preserve their cultures when there are no boundaries?"

Nowhere is this more apparent than in the burgeoning economies of Asia. The emergence of countries like Singapore, Malaysia, China and South Korea into the global information economy has placed them into direct and daily contact with the foreign values of North America and Europe for the first time. Singapore's National Computer Board has launched an ambitious effort to deploy the Internet throughout the country, as part of its IT2000 plan to turn Singapore into an 'intelligent island.' The goal is to bring the Net to every desktop and every home on the island. The country is already one of the most wired places in Asia. One Singaporean in 30 has access through one of the commercial Internet service providers, and many others surf from government or university systems. The NCB plans to turn Singapore's cable television network into one big national intranet, bringing 86 percent of the population on-line by the turn of the century.

Nevertheless, the NCB's concerns are more than technological. Singapore is a cosmopolitan multicultural island, and the government's cultural policies have historically sought to promote a strong sense of common values. The IT2000 project is largely intended to foster a distinct national identity among the various ethnic groups that make up the country's population. "The desire to promote a distinctive culture and society is very much based on the need to maintain social harmony here," said Hao Xiaoming, of the country's Nanyang Technological University. "One way to do this is to promote the Internet not only as an international link but also as a local network to link up the island itself and make internal communication more convenient." However, the Net also threatens to bring a chord of cultural dissonance to the island, in the form of foreign culture and values. By focusing its efforts on creating local content, Singapore hopes to gird itself to meet the threat of Western influences. "Western values are not all that invincible," Hao said. "The most important thing is to what degree people still cling to their own values. If they believe their values are better or equally good compared with Western values, they will not be overwhelmed."

One imported idea that has already created concern among Singapore's Internet planners is the Net's tradition of free access to information. As with other media in Singapore, the Internet is subject to stringent laws governing speech and expression. "It's not to control, but to protect the citizens of Singapore," said Ernie Hai, coordinator of the NCB's Government Internet Project. "In our society, you can state your views, but they have to be correct." Consequently, for all of its technological fervor, Singapore is cautiously implementing the Internet from the top down. Growing numbers of civil servants are coming on-line, and almost every government department already has its own Web site. But the Net has brought the rest of the world to the island's shores, and the NCB actively protects its citizens by routinely blocking access to overseas sites that carry pornographic or otherwise "unacceptable" content.

The Net exposes Singapore to the whole world. But while recognizing that the technological clock cannot be turned back, and that access to the Net is an essential prerequesite for participation in the global economy, the NCB wants some assurances that Singapore won't be drowned in foreign influences. Put bluntly, its goal is to sort through the flood of foreign values and try to block the influences it considers harmful. As alien as this policy may be to North American Internet users accustomed to free access to just about anything at a mouse click, it is symptomatic of the tremors that inevitably erupt when the Net obliterates the political boundaries that once kept diverse cultures comfortably separated.

It's not a question of repression versus freedom, but of different social values and cultural histories. Singapore's cultural values are a manifestation of centuries of social evolution. North Americans, who have not shared the country's unique history, tend to find it all a bit mystifying. Singaporean society—a complex mosaic of many diverse ethnic groups—prizes stability and social order above all, yet Westerners, whose values are rooted in culturally homogeneous communities based on the traditions of common law, can't make sense of how a country that considers itself to be a free society can also be so repressive. In the West, freedom is indivisible, while in Singapore, it must sometimes give way to the interests of social order. The cultures have widely divergent values, goals and sore points, and while friction between them has been mitigated by geography, the collapse of the distinction between margin and centre has brought these cultures into immediate contact.

VI. A Distinct Society

For Canadians the politics of culture have become as stale as week-old hockey scores. For the last two decades the entire constitutional and political life of the country has revolved around the cultural insecurity of the province of Quebec. The vast majority of the province's seven million citizens are native French speakers, sharing a unique cultural history that reaches back more than three centuries to the first French colonies in North America. That distinct character makes Quebec an anomaly in overwhelmingly English-speaking North America, and Quebecers' desire to protect and preserve their language and heritage in an ocean of English is the single dominant issue in the province's politics. The Parti Québécois, whose principal aim is the independence of Quebec from Canada, first formed the provincial government in 1976. Despite the defeat of its independence option in province-wide referendums in 1980, and again in 1995, the government has passed laws to strictly enforce the use of the French language and shore up the province's culture.

Everything from the language used in businesses to the size of English text relative to French on commercial signs is strictly controlled. Government inspectors, charged with the task of enforcing the province's linguistic purity, patrol city streets and suburban malls, looking for infractions of the province's language laws. In September 1995 Microsoft felt the full force of the province's cultural fury when the arrival of the French-language version of its Windows 95 operating system was delayed for a few weeks due to packaging problems.

The following year Quebec passed an amendment to its language law explicitly requiring that "all computer software, including game software and operating systems, whether installed or uninstalled, must be available in French unless no French version exists." If a program exists in French *anywhere* in the world, the same program in English may not appear on the Quebec market, unless the French version is made available simultaneously. The law goes further, and states that anyone who markets software in contravention of the law—from software company executives to shop owners—will be liable to fines under the language charter. Even a computer dealer who sells a system with contraband software pre-installed would be subject to punishment—although he may not know which programs are on the computer.

The reality of the Quebec market is that most software dealers go out of their way to stock French-language versions of programs when they're available. They can't afford to alienate potential consumers in a business as competitive as this, and while many francophone users are content to buy English software, providing products in the language of the majority of consumers is just good business. Moreover, high-technology companies have been repeatedly recognized by the government as good corporate citizens in terms of language. IBM Canada was even awarded a Blue Seal by the Office de la Langue Française (the government department responsible for enforcement of the provincial language laws) a decade ago to recognize its compliance with the language charter. When it comes to culture the government of Quebec is willing to sacrifice any ally in the pursuit of linguistic purity. The information revolution is, above all, the occasion for even more exacting cultural standards.

The Internet is explicitly a cultural issue in Quebec. "Everything is a cultural issue in Quebec," said Louise Beaudoin, the provincial Minister of Culture and Communications. Quebec's Information Highway Secretariat, the body responsible for establishing and implementing the province's information technology policies, is a department in her ministry, which is also, incidentally, responsible for enforcing the province's language and culture laws. Beaudoin likes to talk about the 'opportunity' the new technologies offer for the Parti Québécois' 'social project,' but at the bottom of it all is fear. Language equals culture in Quebec, and every policy must be balanced against the real or perceived possibility that the French language could be swamped in the overwhelmingly English datastream. The official statement of the province's information highway policy, released in 1995, made that abundantly clear.

> In a knowledge-based economy, language is the principal concern. In Quebec, the preservation of the French language, in so far as it is a reflection and expression of the culture is the heart of these preoccupations. For many observers, the future of the French language depends in large part on its capacity to take the route of the information highway.[4] [Translation by the author]

The task—the francization of the Net—has the status of a holy crusade in Quebec. "We're two percent of the population in North America, and this is a profound insecurity," Beaudoin said. "I don't believe that if everything is in English on the Net, that everyone will be happier everywhere in the world—because this means homogenization."

Quebec's historical insecurity has been greatly amplified by the information revolution's tendency to erode the geopolitical divisions between countries and cultures. The perceived threat posed by the rest of the world—or at least the rest of North America—figured prominently in the government's very first statement on information highway policy. The government's 1995 policy statement stressed the threat faced by Quebec, "isolated in a continent of 250 million English-speakers."[5] It observed that by subverting geographical distance the Net has brought the outsiders to Quebec's doors. This sense of being a threatened enclave in the midst of foreign hordes has informed the province's information highway policy ever since.

A series of government reports sounded the alarm about the threat that an English Internet might have on the French culture of Quebec. A recent working document prepared by an all-party committee of the provincial legislature warned that the use of English as the Net's lingua franca threatens to marginalize all other languages in the emerging global economy. While conceding the obvious—that the Internet is an international medium, that most of the new technologies have been developed in the United States, that the Net had its origin there, and that there are more American users on-line than any other nationality—the committee was nevertheless alarmed that English was being used more than French! The first language of Quebec is, at best, a second language on the Internet, and in the minds of the committee members, that means that French risks being marginalized out of existence in the information age.[6] Quebec is vastly overreacting. There is a huge difference between adopting English as a convenient second language to communicate with the rest of the world, and allowing it to replace your mother tongue. "From the point of view of someone who's trying to protect French in Quebec, I can understand that the Internet looks like a big threat, because it's all English," Anderson said. "But I can tell you as a native speaker of English that what's going on there is not real English. You especially see this if you look at foreigners who are using English, because they are using it as a second language." The same report concluded that 91 percent of the information on the Internet is in English, but failed to point out exactly *how* Quebecers are to communicate with users from the United States, Britain and the rest of Canada *without* using English.

The Quebec government has implemented an aggressive program to develop French language Internet technologies and content. "Our view is that it is important for the information highway to translate the realities of the world," said Robert Thivierge, the deputy minister responsible for the Information Highway Secretariat. "There are several linguistic groups in the world, and one is French. It is important that reality is represented on the information highway." To do that the secretariat recommended that the government encourage the development of home-grown, on-line content in the French language. Thivierge contends that the key to raising the province's cultural profile is for Quebecers to provide the content that the rest of the world wants to see. The government is intent on making Quebec's language and culture the focus of its information highway strategy.

One of the peculiarities of Quebec's political culture is its long tradition of corporatism. Perhaps as a response to the perception that Quebecers are a distinct cultural group in North America, the province's leaders have always tried to set policy and express the identity of Quebec through organizations and public assemblies. The individual Quebecer is seen as an atom in the greater whole of the national identity, and not as a separate being with individual interests and aspirations. The nationalist cultural awakening of the 1960s and 1970s was characterized, above all, by grand events—shows, festivals, demonstrations—that mobilized the culture en masse, rather than having it expressed in the daily lives of the people as individuals. Quebec's Internet policy is simply an expression of this cultural corporatism, and there lies the problem. The medium's technological and social infrastructure effectively subverts the centralized, mass cultural manifestations that Quebec's nationalist leaders have so long relied on to mobilize the public. The user creates the content and experience, rather than having it directed by some higher power, and Quebec is having trouble coming to terms with that.

The Information Highway Secretariat awarded $10 million in subsidies to Quebec developers in 1996, with priority going to those who promoted the on-line use of the French language. While Beaudoin admits that this policy discriminates against English-speakers who make up one fifth of the province's population, she maintains it is a necessary evil, and equates it with affirmative action programs for women and minorities. It's a typical Quebec conundrum: the government can't seem to decide whether French Quebec is a minority or a majority. Its rhetoric always stresses the French dominance within Quebec, but it employs policies appropriate to a disadvantaged minority. In Quebec the real or more often imagined historical oppression of Quebec is the key to everything. Ironically, nationalists like Beaudoin and Thivierge seek to establish dominance by assuming the status of a threatened minority, with the ultimate goal of forging a national mono-culture to the exclusion of everyone else. The attitude seems to be that if Quebec's multilingual and multicultural reality doesn't quite coincide with that image, then everything that isn't French should be hidden away.

In terms of the Internet, this strategy may be neither effective nor necessary. The means and the motive are mismatched, resulting in an absurd industrial and technological policy. While it's hard to argue with the government's desire to promote the unique, vibrant culture of the province, it is being done at the expense of Quebec's competitiveness in the information economy and the technological enfranchisement of its citizens. More to the point, Quebec has put very little effort into getting the *people* of Quebec on-line. A Free-Net project in Montreal languished and died largely because the government never gave it funding priority. Though other factors, including organizational problems and personality conflicts, were partly to blame, the government's inability to commit itself on the funding question for *three years* was the project's death warrant. Despite overheated populist rhetoric, Quebec has shown no interest in encouraging public Internet use.

There has been an all-out effort to connect public institutions like hospitals and libraries to the Net, and to encourage corporate content developers, but the ordinary citizen has been left out. "What's important is the information that's carried on it," said Thivierge, "and not the wires." Unfortunately, until those wires are actually connected, and there are people to use it, whatever information they may carry is irrelevant. The province's information infrastructure simply isn't advanced enough yet for the information highway to be anything but a technological issue, and what language the wires carry should be incidental to the technology itself. Moreover, only four percent of the homes in the province have Internet access—one of the lowest rate in North America—so whatever the language and content of the Net happens to be is pretty much irrelevant. It's as if the government is throwing a reveillon—a traditional Québécois New Year's celebration—that no one can attend!

The official line is that by encouraging the development of French-language content the government of Quebec is providing Quebecers with an incentive to *use* the Net, forgetting that the best content in the world is meaningless without users. But that appears to be the point for now. What concerns the government most is not the survival of the French language and culture in Quebec, but its control over it. After decades of building up a French mono-culture in Quebec, and denying that any other cultural group even exists in the province, the gatekeepers are suddenly confronted with the heterogeneous culture of the Net. English is undeniably the Net's lingua franca, but that clearly doesn't mean the same thing as cultural homogeneity. In Quebec at least, English has become a battleground for cultural heterogeneity.

The creation of distinct on-line communities has become a defining element of the cosmopolitan Internet culture. "People don't go onto the Internet to find someone to argue with, they gone on to find somebody to agree with—they go on the Internet to find home," Anderson said. "The typical feature of the Internet is creolizations. What you get is mixed languages, you get people who are themselves mixtures and the reason you get this is the folks who write the rules in other places are absent." It's the Internet's lack of conventional cultural boundaries, and not the dominance of English that most concerns cultural gatekeepers like Beaudoin. It's their relevance, not their culture, that is threatened. They intuitively feel that their authority is being challenged by the information revolution. In Quebec, culture, or more precisely, cultural boundaries inform everything... but the Net has no boundaries.

VII. Digital Diasporas

For many of the world's minority cultures the absence of geographical boundaries is the Net's greatest value. Cultural minorities that don't have the advantage of borders and governments to promote their unique identities and languages have begun to turn to the Internet. Small groups from East Timor to the Frisian Islands whose languages and traditions are often suppressed by dominant majorities, or whose members are scattered in exile communities around the world, are literally blossoming on-line. The Net is a medium of expression for diaspora cultures, regardless of geographical distance. It's an environment where the experience of diaspora can be reinvented every day, because the difference between margin and centre is moot.

That has been the experience of the Welsh community in North America and Wales. The creation of an on-line community of Welsh speakers, centered on the Welsh-L listserv, followed quickly on the heels of a cultural renaissance of the 1970s and 1980s, which saw the establishment of Welsh-language broadcasting services in Great Britain, and the explosive growth of schools offering instruction in the national language. "If there is no Welsh language, there is no Welsh culture," said Pawl Birt, a professor of Celtic languages at the University of Ottawa and one-time president of Cymdeithas Madog, the Welsh studies institute of North America. "Our language is the oxygen that feeds every aspect of our culture."

After two centuries of emigration and many more centuries of English cultural domination, the Welsh community outside Wales was in danger of losing touch with the homeland. However, they have succeeded in recreating a virtual homeland on the Net. The Internet is thick with Welsh sites, from language courses and the Welsh Language Board, to rugby teams and information on the Welsh rock scene. Most sites are bilingual—Welsh and English—but a good number of them are Welsh only. "The Internet gives you a buoyancy, it makes you feel like part of a cultural continuum," said Birt, who left his native Wales for Canada eight years ago. "Welsh speakers feel much less isolated, like they're part of Wales. The Internet may ultimately change the traditional view of Welsh culture, and it may not be too much to start speaking of a cyber-Welsh society."

This isn't an isolated experience. At the University of California in Santa Barbara, Internet culture watcher Alan Liu has catalogued a vast number of minority culture Internet resources from around the world. He believes that the Net may be both the agent and stage for a revolution that can only invigorate minority cultures. "The Internet is not a middle-man-controlled publishing medium, so for the first time we have a vehicle that allows these groups to define themselves," he said. "Because the barrier to entry is so low, it's no longer the case that some central publisher will decide what a group will be called, or how it will be represented. And because it's interactive, the medium gives members of these cultures a virtual place where they can express their identity, no matter where they are geographically."

The Internet is not one culture, but many. Whether it remains that way or evolves into a commercial mono-culture remains to be seen. The danger is that in the rush to profit from the information revolution, groups like the Welsh could see their traditions commercialized as commercial curios for yuppie consumers in North America.

The future of minority cultures on the Net could go any of three ways. It could revert to the anarchic hetero-culture of the early days of the Net, with no unifying foundation but the Net itself. It may evolve into a settled multicultural mosaic, or, most ominously, become a mono-culture with the illusion of multiculturalism. In the last scenario, culture becomes little more than just another information commodity to be traded in the corporate global economy. Ethnicity could be neatly packaged so that it's only another product on the shelf of the multicultural supermarket. That has already happened with the world's musical traditions and ethnic cuisines, and it is the ultimate fate of culture and identity when they are separated—as Quebec is doing—from the *people* who create it.

There is a real urgency in the need for cultural groups to appropriate and internalize the tools of the information age, and equally to extend a digital franchise to their members. The technology must 'go native' before local cultures are commodified into irrelevance. Many groups like the Welsh have had admirable successes, but with the growth of the global information economy, the explosion of Internet use and the irresistible pace of innovation, the world's culture may be rapidly approaching a critical juncture. The thing to be feared is not language and foreign influences, but the direction that the Net is moving in—toward more demanding technical requirements and corporate control. In a practical sense, the only way for the world's minority cultures—and on the net, all cultures that aren't American are minorities—to survive the tide of American art, entertainment, values and language, is to build their own identities on-line, person by person. They have to take control of the technologies of information from the corporate economy and make them their own.

CHAPTER FOUR

Politics by Other Means

I. The Streets of Belgrade

BELGRADE WAS A TINDERBOX in the cold, grey Balkan autumn of 1996. On November 17 opposition parties loosely united in the Zajedno coalition had won a majority of seats in Serbia-wide municipal elections. The ruling Socialist Party initially conceded defeat, but before the dust had settled on November 18, president Solobodan Milosevic annulled the results, citing widespread voting irregularities. Within days, tens of thousand of protesters had begun to march in the Serbian capital's streets, denouncing the president and calling on the international community to force him to reinstate the results.

The protests grew by the day, and by the beginning of December hundreds of thousands of marchers, bearing Serbian flags, placards and effigies of Milosevic regularly crowded Belgrade's streets. For all the slogans, chants and cries it was a strangely silent protest. Almost all the newspapers, television and radio stations in the capital are organs of the governing party, and to judge by their coverage, Belgrade was a peaceful and quiet city during those weeks when autumn gave way to winter. Dragan Tomic, speaker of the Serbian parliament, denounced the demonstrators as proto-fascists, and likened their leaders to Nazis, an unlikely lie that might have carried the weight of truth had two radio stations not stood against the party line. Radio B92 and Radio Index, the only independent media outlets in the country, became the unofficial focal point of the Zajedno campaign, broadcasting up-to-the-minute reports on the demonstrations, and serving as a clearinghouse of information about pro-democracy activities.

It was a situation that Milosevic's government could not long abide, and in the last week of November it began jamming the stations' signals. Though they could only be heard in central Belgrade the message still managed to get out by fax and telephone. One Serbian Internet user told of holding his telephone receiver up to the radio so his out-of-town friends could hear the B92 news. On December 3 the Serbian government had finally had enough and revoked the stations' licences and forced them off the air, hoping that they would now be silenced once and for all. However, nothing is that simple in the information age. In the tense weeks before their transmitters were shut down, B92 and Radio Index had appeared on the Net. At first their sites—hosted in the Netherlands—carried press releases and short reports on each day's protests, but by December, B92 began providing digitized audio of its daily news broadcasts in both English and Serbian. Belgrade residents couldn't tune into their favourite station by radio, but millions of Internet users in Serbia and around the world could listen to the twice-daily updates on the Net.

It's impossible to say exactly what role these Internet broadcasts played in the anti-government struggle, but by the middle of December the Milosevic government had begun to feel increasing pressure from the United States, the European Union and even Montenegro, its neighbor in the Yugoslav federation. Already suffering the effects of international trade sanctions for its support of Serbian nationalists in Bosnia, Serbia was in an untenable position and, as the daily number of protesters rose to a quarter of a million, Milosevic began to give ground. On December 14 a Serbian court upheld the November 17 election results in one municipality, and others began to follow. Significantly, at the same time, the government lifted the ban on B92, whose new slogan was "you know you're really independent when everybody hates you," and Radio Index.

B92's Internet experiment was a dramatic confirmation of the new political realities in the information age. Less than a decade ago, it took little effort for repressive regimes to silence dissenting media. Milosevic would only have had to send his secret police to smash opposition newspaper presses and turn off the power to radio and television stations. The *means* of producing and distributing information were neatly contained behind geographical borders. The information revolution has changed all that. As Negroponte has argued, the bits of digital information know no borders, and short of confiscating every computer and modem in the country and cutting off all international telephone lines, there is very little a repressive government can do to stem the datastream. The new reality of politics in the information age is that private internal affairs are now distressingly public and that, on the Internet, the whole world is watching.

II. Digital Underground

It has become an axiom of the information revolution that the new technologies will soon fundamentally change the nature of politics around the world. Some of the more utopian commentators have suggested that we are poised on the brink of the dissolution of the state, that interactive village meetings will soon replace the venerable and unresponsive forms of representative, electoral democracy. Yet, at least at the centre of the American heartland of the information revolution, there is little evidence that the new technologies will bring much change to traditional, pork-barrel politicking.

Mainstream politics came to the Internet in a big way in the run-up to the 1996 American presidential election. We knew the Net had arrived when even the stodgiest geriatric political hopefuls—like then-Senator Bob Dole—had a web page. With minor exceptions, everyone who wanted to be president was already on-line by the time the primaries shifted into high gear in the spring. Surfers were invited to calculate how the flat tax would affect them at Steve Forbes' page and read Lamar Alexander's anti-everything-yet-too-liberal-for-Buchanan message at the Tennessee hopeful's site. At Dole's own site, we were invited to download Bob Dole screen savers and desktop wallpaper. Presumably, Dole's true believers needed to have the old goat's face staring out from their computer screens at all hours of the day or night for inspiration—the digital version of a dashboard Jesus. Significantly, one dissenting user became a kind of information age heckler by setting up a site with a near-identical address and appearance that diverted legions of would-be supporters to a withering satire of Dole's political program.

It was all business-as-usual. Even at the home page maintained by ultra-conservative candidate Pat Buchanan—both the most revealing, and strangely the most appropriate of the candidates' on-line storefronts for the Internet environment—traditional, rather than technological revolution was the dominant message. A sepia-toned welcome screen, complete with turn-of-the-century swashes and flourishes, announced that *this* was the home of good, old-fashioned values. Buchanan is pro-life, pro-family, and pro-American, of course, but his Web site made it abundantly clear that he's anti-everything else. The candidate's negative message was reinforced throughout by militaristic imagery. Good, clean-living, *Christian* Americans are at war, and the "Buchanan Brigade" is only one unit in the fundamentalists' army of light. Buchanan made the text of endorsements from religious extremists available to all who visited the site. Significantly there was no mention of the links to white supremacist and neo-fascist groups that had forced two of his senior aides to resign midway through the primary process.

The on-line primary campaign was met by a torrent of dissenting voices and serious critique almost from the day the first hopeful threw his virtual hat into the ring, but most of this came from traditional sources. For example, *Congressional Quarterly,* which has been covering Washington politics in print for decades, used its considerable resources to bring campaign profiles and biographies of the candidates—including some choice bits of information they'd like to keep hidden—to the Net. Elsewhere otherwise marginal political organizations, activists and private citizens published detailed analyses of the candidates' policies, criticized the system itself and debated the issues. Considered dissent was nothing new to the Net; indeed, it is a deeply ingrained part of the on-line society. The mainstream political process attempted to appropriate the Net as another medium for its slogans and platforms, ignoring the fact that this new medium—this new social and political space—stands in glaring opposition to its basic forms and assumptions. The Net is a political environment, but it is also a politically subversive technology, enabling a discourse that promises to redefine traditional civil society.

New information technologies have a funny way of subverting the established order. When the printing press was invented, books—particularly the Bible—became readily available to the general public. With that came literacy, and when people actually started *reading* the Bible, they started asking questions about hitherto immutable truths. They started a reformation. Newspapers and political pamphlets, printed with a new technology of cheap moveable type, let Thomas Paine and Georges Danton inaugurate an age of revolution. Television coverage of the war in Vietnam galvanized the American public against the conflict in a way no protest march or riot could equal. The true face of the war—the *information* of the war—flowed into the living rooms and dens of ordinary Americans with such speed and abandon that earlier governments would have called it treason. It was no wonder that, a generation later in the Persian Gulf, the military media masters' first order of business was to strictly limit what sounds and images could be broadcast on television.

Information technologies are subversive because they offer potential enfranchisement to the disenfranchised, and voice to the voiceless. And the Internet, with its elimination of the distinctions of margin and centre promises to bring the fringe into the mainstream. Parties and factions march, countermarch and rally in the metaphoric place that is the social space of the Net. One way to look at the Internet is as civil society, a place of social congregation that "is active precisely at those moments when, and in those locales where, people have gathered."[1] It is that environment between private life and state, where citizens engage in the political discourse with government and political institutions.

It is as fruitless to search for a single politics of the Internet as it would be to listen for one pure voice in the happy-hour din of a downtown bar. The Net is a conversation of millions of voices, and its politics is a weave of countless threads. Indeed, the very political character of the Net is anarchic, which is not to say that it monolithically espouses an ideology of anarchism—though more than a few users certainly profess to—but that of the hundreds of individual discourses on-line, no one is dominant. To a certain extent, this political polyphony is a reflection of the redundant and non-centralized communications structure. Also built into the network infrastructure "was high-speed communications and as much openness as possible, because those are the values of the scientists and engineers who designed it," said Jon Anderson.

Moreover, the values of the engineers and developers who created the tools and infrastructures of the information revolution—immortalized by Steven Levy in *Hackers* as 'The Hacker Ethic'—became the foundation of the Net's political landscape. The traditional hacker dictum that all information should be free has been a guiding principle of the Internet since the days when it was the preserve of scientists and graduate students. However, in the wake of the information revolution, something extraordinary happened. What had been non-conformism to the engineers and hackers who had pioneered the Net soon became politics. The one unifying factor of the multitude of on-line discourses is this resistance to any force or agency that would impede information. If all information should be free, then it follows that anything impeding its freedom should be resisted. Consequently, the Hacker Ethic is a kind of political resistance to any centralized or corporate control of information.

The community that began to coalesce around the Internet and computer bulletin boards in the 1980s had the appearance of a political counterculture, espousing common values and ideals. Writing of the Well, an early bulletin board system in the San Francisco area, Howard Rheingold perceives the information revolution as a continuation of 1960s-vintage Hippie idealism, and the on-line community as an extension of the flower-power counterculture.[2] To veterans of the Summer of Love and Woodstock, it looked like the 1960s all over again, but such a facile comparison ignores the fact that the on-line counterculture built on the hacker ethic is a sharp departure from the youth movement of a generation ago...

> ... a shift in the relation of countercultural activity to technology, a shift in which a software-based technoculture, organized around outlawed libertarian principles about free access to information and communication, has come to replace a dissenting culture organized around the demonizing of abject hardware structures.[3]

Moreover, this counterculture—such as it is—has little ideological or demographic cohesion. While most members of the on-line community will at least offer lip-service to the ideals of the hacker ethic and share a vague libertarian inclination and mistrust of authority, *that is about as far as it goes.* There is no unified countercultural politics on the Net.

Technologically the Net is a radically decentralized structure, which makes it extraordinarily difficult for one group to dominate the discourse. Conversely, it encourages the information enfranchisement of a wide spectrum of ideologies and opinions. While the mainstream political discourse is circumscribed by the institutions of government with the ability to disseminate ideas and information through the mass media, and authority based largely on economic standards of status and prestige, the on-line discourse is open and interactive. It circumvents traditional modes of authority and information control. Thus the debate isn't limited to liberals, conservatives, Republicans and Democrats; a full spectrum of ideas from anarchists to born-again fascists—and everything in between—can be heard at equal volume, and they have as much credibility as their fellow users choose to give them.

It's impossible to say when politics first appeared on-line, but it's safe to assume that it happened fairly early. The Internet was conceived and built within the context of university study and research, at a time when one could hardly be a student or professor without at least *appearing* to be committed to one cause or another. It didn't take long for individual activists and organizations to discover that they could use the new media to network among themselves and get their message out to the world at large. PeaceNet and the Association for Progressive Communications were certainly among the first on-line activist organizations, with the latter coming on the Fidonet BBS network as early as 1982. APC has cultivated links in many countries over the years and was, for a time, one of the few sources of accurate information available on the Nicaraguan Revolution for those of us in North America. Both organizations continue to be clearinghouses of information on progressive and left-wing issues as diverse as women's rights, the environment, prison reform and human rights. Getting access to APC and PeaceNet was the reason why many Latin American solidarity groups, student newspapers and campus radio stations acquired computers in the first place in the early-1980s.

For all of their exertions, on-line activists are mainly propagandists, focusing on *getting the message out* and communicating among themselves. As one veteran activist wryly commented, "people in a newsgroup may be in Australia or Zimbabwe, which is way cool, but not helpful to getting folks out in front of the Town Hall of Topeka next Friday night."[4] Nevertheless, many activists have found that, in a world where information and power are more closely connected than ever before, it may not be necessary to hold a demonstration on the steps of Topeka's Town Hall when you have the tools to shame your opponents—whether they are government or corporate—with the truth. After all, the political activism of the information age is the politics of information.

III. Basta!

New Year's Day 1994 was supposed to be the date of Mexico's coming-out party. Long the poor, feeble sibling of North America's powerful industrial economies, the North American Free Trade Agreement, which came into effect at midnight, seemed to mark Mexico's entry into the neo-liberal fraternity as a mature, developed nation. The global free market had crossed the Rio Grande and for the Mexican government and business leaders, NAFTA was proof that they had arrived.

In the early hours of the year, armed men emerged from the jungles of Chiapas in southern Mexico and quickly occupied the towns of San Cristobal, Altamirano, Ocosingo and Las Margaritas. The masked guerrillas threw open the jails and government buildings and issued the Declaration of War against the Mexican government and foreign business interests that would resonate throughout the country.

> We have been denied the most elemental preparation so they can use us as cannon fodder and pillage the wealth of our country. They don't care that we have nothing, absolutely nothing, not even a roof over our heads, no land, no work, no health care, no food nor education. Nor are we able to freely and democratically elect our political representatives, nor is there independence from foreigners, nor is there peace nor justice for ourselves and our children.[5]

Though it began as a peasant rebellion led by the Zapatista National Liberation Army—the Ejército Zapatista de Liberación Nacional or EZLN—in the name of the indigenous people of Chiapas, the struggle had broader overtones from the very beginning. It was no accident that the EZLN emerged from the jungles on the first day of NAFTA. The Declaration of War clearly identified foreign interests and the selling-out of Mexico's land and resources as the heart of their grievances. The EZLN's first concern was the Mayan people of Chiapas who lived in a desperate state of grinding poverty, exploited by the government and landowners alike. It was clear from the outset that the struggle went far beyond local needs.

> But today, we say *enough is enough*. We are the inheritors of the true builders of our nation. The dispossessed, we are millions and we thereby call upon our brothers and sisters to join this struggle as the only path, so that we will not die of hunger due to the insatiable ambition of a 70 year dictatorship led by a clique of traitors that represent the most conservative and sell-out groups.[6]

For the Mexican government the uprising was an embarrassment. The country had just been welcomed into an association with the United States and Canada, two of the world's most advanced economies, as an equal! And yet, here was a band of masked and armed peasants calling attention to the fact that, despite the efforts of the Salinas government to hide it, Mexico remained a corrupt, feudal society under the economic thrall of its northern neighbours.

Not surprisingly, the government's first response was to attack. On January 2, 1994 police and federal troops moved into Zapatista-controlled territory, supported by armoured vehicles and aircraft. The counter-offensive was initially successful in military terms. Within a week, the EZLN had been thrown from most of the land it had occupied on January 1. It was a political and economic disaster for the Mexican government. Public support for the rebels grew throughout the country as revelations of military atrocities in Chiapas emerged. Despite the brand-new Free Trade Agreement, the Mexican stock market and currency began a long and ruinous slide. On January 12 the government declared a cease-fire in Chiapas, and by the end of the month had agreed to discuss the Zapatista grievances.

Talks between the EZLN and the Mexican government never produced concrete results, despite the latter's insistence that it would take the Zapatistas' demands seriously. Those demands called for a reform of Mexico's de facto one-party political system that has historically been propped up by widespread corruption and patronage; social, economic and land reforms throughout Mexico—including the repeal of NAFTA—and the de-militarization of Chiapas. The government would not give ground on the political questions, and offered a commitment to develop health and educational resources only in the part of Chiapas directly affected by the uprising. On June 10 the Zapatista General Command (CCRI-CG of the EZLN) flatly rejected the offer. Their struggle was not for short-term, regionally-specific goals, but for a basic transformation of Mexico as a whole. In their communiqué the Zapatistas lashed out at what they perceived to be the one-party rule of the Institutional Revolutionary Party in Mexico, and condemned the government's offers as little more than a smoke screen designed to shore up its position. The bottom line of the Zapatista struggle was democracy in Mexico, and the rebels maintained that nothing less than the death of the current Mexican political system was its precondition. "There will be no real solutions until the situation in Mexico as a whole is resolved."[7]

The government's reaction was to ignore the EZLN and hope it would just go away, but it didn't. The Zapatistas' demands resonated throughout Mexico, as revelations of government corruption followed on the near-disintegration of the economy, ultimately inspiring similar uprisings in other impoverished regions of the country and leading to the birth of a national political movement that would eventually call itself the Zapatista National Liberation Front—the Frente Zapatista de Liberación Nacional, or FZLN. A bloody government offensive in the winter of 1995 failed to destroy the EZLN and only succeeded in increasing its support throughout the country and around the world.

Prior to 1994 few Americans or Canadians were aware that there was a place called Chiapas, or that the Maya were anything more than an archaeological curiosity. The Mexican government's efforts to contain peasant unrest in the region had been largely successful for decades. Heavily armed troops patrolled the roads and villages, and the state-run broadcasting network kept the lid tightly clamped on whatever trickles of information might leak out to the rest of the world. Only a decade ago, that would have been it. The Mexican army would have been able to use its numerical and material superiority to quietly crush the revolution, round up its leaders, and restore the authority of the central government. But nothing is quiet in the information age. A principal aim of the Mexican government campaign was the isolation of the EZLN, so that it could either be destroyed by force of arms or co-opted out of existence when the rebels' leadership was exhausted by tilting at immovable windmills. But the Zapatistas quickly turned their struggle into a war of ideas and words, calling on the world to help, or at the very least, to witness what was happening in the jungles of Chiapas. The Zapatistas blew the government-imposed lid right off with a unique and revolutionary combination of direct military action, carefully choreographed political theatre and an information campaign unlike anything attempted before.

The EZLN and its supporters in the United States, Canada and Europe is not simply a traditional guerrilla army with jungle hideouts, assault rifles and romantic noms-de-guerre. They are the first of a new breed of information guerrillas, fighting their war in the media and on the Internet as much as on the barricades. Other political uprisings, notably the Chinese pro-democracy movement, have used e-mail and faxes in the past to get around censors and speak to the world, but the Zapatistas are something new, and the tools of the information revolution have been woven into the very fabric of their social and political revolution.

"On a couple of different levels, it's a knowledge and information-based struggle," said Justin Paulson, an American university student who manages the EZLN's semi-official Web site from Swarthmore College. "There's the basic attempt to break the hegemony of concepts that hold that neo-liberalism is the natural economic order. We see that even more in the first world; in Mexico, people don't even believe that anymore. They were believing it, to a certain extent, under Carlos Salinas, but now the government has a very low popularity, and nobody believes its economic policies anymore." The Zapatistas have been able to get their communiqués and tracts into print all over the world. They have been able to present an incisive critique of neo-Liberalism, and a socio-economic alternative that boils down to the environmentalist creed of think globally and act locally. Moreover, the EZLN's deft use of the technologies of the information revolution has demonstrated repeatedly that it can get its message past the extremely partisan Mexican press to the world. There's a sense that the government can't hide anything, that the whole world is watching.

Paulson is one of thousands of Zapatista supporters in Mexico and around the world who have helped the EZLN move their struggle from Chiapas to the world stage. The revolution came to the Net almost as soon as the Zapatistas emerged from the Lacandona jungle on New Year's Day 1994. Subcomandante Marcos, the EZLN's charismatic spokesman, composes his communiqués on a laptop computer which are then printed and smuggled to San Cristobal to be faxed and e-mailed to media outlets and Zapatista supporters around the world. "This is a different kind of revolution," said the University of Texas' Harry Cleaver. An EZLN supporter, he has observed the events in Chiapas with intense interest. "The opposition to the American intervention in Central America [in the 1980s] was very similar to the opposition to the American war in Vietnam. It was more of an opposition to the United States going in and taking over other people's countries and governments and killing lots of people, than it was support for the particular groups that were in revolt. There's something different about the Zapatistas. They're not looking for the seizure of power. They're not a Leninist organization even though they may have come out of one a long time ago. What they have to say and the way they have to say it is very different from anything that has come before, and speaks new kinds of truths."

Though the revolution in Chiapas is itself a military struggle, the armed component serves principally to protect the EZLN's membership against arrest and to advance its agenda throughout Mexico. Localized military success in Chiapas would do nothing to reform the Mexican state and economy, so through the Frente and the Internet the struggle has become a campaign of ideas and information. Both the strategy and organization of the EZLN are ideally suited to this kind of revolution. The rebels have established a decentralized political discourse that dovetails perfectly with the social and technological infrastructure of the Net. Their strategy has been to link their struggle to grassroots movements and social reformers around the world. In effect the EZLN has managed to create a global network of ideas that mirrors the modes of communication and community implicit in the Internet. In one sense the Zapatistas are just one of many political movements and communities in an environment characterized by competing, but sometimes complementary, discourses. It's also a fertile environment for a political movement whose organization and message are predicated on global linkages of common cause between local struggles.[8] Those links have been nurtured through conferences and forums in Mexico and elsewhere—the EZLN's 1997 international conference was scheduled to take place in Berlin—and on the Net.

Though Paulson is cautious to point out that the war in Chiapas is a very real war with real death and privation, he can't play down the importance of the new medium. "The Internet is important—it's a newspaper you can read really quickly, you don't have to wait for its publication," he said. "It's a way that you can spread information so you can mobilize people. It's a technique that you can force your opponent in the struggle to not put forward a policy of 'the public will forget about you.'" In this case, the Zapatistas have stayed very public, both in Mexico and on the international stage, while hiding from the might of the Mexican army. When both sides sit down to the bargaining table, the government is consequently forced to deal openly and, presumably, in good faith. The EZLN may not have the *whole* world behind it, but the Mexican government, starved for foreign investment and international respect, can't ignore the support the rebels do have.

Perhaps most importantly the Net has allowed the Zapatistas to control their presentation in the mainstream American media. In the past the mainstream media's corporate bias has adversely affected the portrayal of Latin American national liberation movements to the American people. The Net has eroded their monopoly on information, with the result that the EZLN has typically been portrayed more sympathetically than any similar organization in the past. "The Zapatistas knew exactly what the press was going to be doing, and what the government was going to be doing," Paulson explained "At the very beginning, in their communiqués, they said 'you will call us this... you will call us that... and you'll be wrong on all counts.' They started off, at the very beginning, pre-empting all the negative characterizations that tend to go along with media coverage of revolutionary struggles. 'You'll call us drug traffickers, you'll call us communists...' And they were able to construct, very early on, their own image of themselves, that took the media by surprise and became what everyone was able to characterize them as. It was silly and reactionary after the first week for anyone to try to call them drug traffickers or dogmatic communists. The media *tried* to say that they were infiltrated by Central American revolutionaries from Honduras or Cuba, but the EZLN explained very clearly why that couldn't be so."

Significantly, the Zapatista struggle is a response to the same economic conditions that attended the first stirrings of the information revolution. It is, in many ways, a revolution *within* a revolution, answering the challenges posed by the emergence of a global, neo-liberal corporate hegemony. The North American Free Trade Agreement was the spark that set the revolution on its course, but even NAFTA is merely a manifestation of the process of globalization that has driven the work economy for the last decade. "The Zapatistas have understood that neo-liberalism is the general character of capitalist policy-making in this period," Cleaver said. What the Mexicans call neo-liberalism is Reaganomics in the United States, Thatcherism in Great Britain, and what the International Monetary Fund calls restructuring or austerity in the developing world. Despite the passing of the great conservative government administrations of the 1980s, their economic policies and ideology—often cloaked as deficit-cutting and fiscal responsibility—continue to dominate the industrial world and the global economy. Through their conferences and information campaign the Zapatistas have been able to demonstrate that neo-Liberal ideology dominates the economic policies of most of the world's governments.

The things that the EZLN is attacking and rejecting turn out to be the things that many people at the grassroots are attacking and rejecting all over the world. The message is clear—despite local concerns, we're all in this together.

Consequently the EZLN has emerged as the leader of the resistance to the neo-liberal hegemony manifested in the expansion of the American-led information economy. Ironically, the new economy's growing global infrastructure is, simultaneously, the most important forum for distributing the opposition's message and ideas. In the wake of the new order's inability to deliver on its promise of a chicken in every pot and a videocassette recorder in every living room the Zapatistas' message has particular resonance. Translated into several languages and widely distributed in electronic form, Subcomandante Marcos' attacks on the neo-liberalism—often in the form of serio-comic platonic dialogues with a jungle beetle named Durito—have become the essential texts of the resistance. "Subcomandante Marcos is already doing a lot of the popularizing of the conflict on his own," Paulson said. "It's basically his lead that we're following, by making it an information revolution to the extent that we're going to publish things and get things published." Significantly, it's not a dogmatic struggle. At a fundamental level the Zapatistas are indigenous people in Chiapas fighting for political and economic liberation. But it's also a global struggle against corporate capitalism, shorn of tiresome Comintern rhetoric. Despite the growing resistance to the ideological legacy of Reagan, Salinas, Mulroney and Thatcher, it took the EZLN and the Frente campaign to knit it all into a coherent global discourse. That could not have been accomplished without the Internet.

With the success of the EZLN, the Frente and the supporters on the Net, it didn't take long for similar revolutionary movements to move onto the Net. In a report prepared for the Office of the Assistant Secretary of Defense for Special Operations and Low-Intensity Conflict, Pentagon researcher Charles Swett observed that

> The Internet has been playing an increasingly important role in international politics. One highly significant effect of Internet use overseas has been to circumvent the informational controls imposed by authoritarian regimes on their citizens.[9]

Whether he felt this was cause for celebration or alarm isn't clear, but his conclusion that individual activists in the developing world, equipped with portable computers and telecommunications technologies are likely to bring the Internet with them is a clear reference to the EZLN. Marcos and his comrades represent the future of national liberation struggles. The Internet is both an opportunity and an imperative for local movements to unite and work more closely than they have in the past. While this new reality of struggle has a lot to do with the emergence of inexpensive and accessible global telecommunications and information technologies, it is also a response to the globalization of ideology and trade in the corporate information economy. There is a growing realization, reflected most clearly in Chiapas, that local resistance must be global resistance, and the Zapatistas are becoming the de facto leaders of this movement against the neo-liberal hegemony.

IV. Cyberwar

It sounds like the title of a science fiction novel or a shoot-'em-up computer game, but infowar or cyberwar is rapidly becoming a reality in the wake of the information revolution. In 1996 the United States government Department of Defense's *Militarily Critical Technologies List*, an annual reckoning of the state of military technology in the U.S., included a section on Information Warfare for the first time.

Around the time that Charles Babbage was sketching the plans for his Difference Engine, the Prussian strategist Karl von Clausewitz observed that war is regarded as nothing but the extension of politics by other means. It could equally be said today that information warfare is the extension of the politics of information. Though cyberwar sounds more like the title of a science fiction novel than the practical continuation of politics in the information age, there is a dawning realization among activists and governments alike that the Net is becoming an essential element in military and paramilitary strategy. The Zapatistas' Internet campaign, though carried out by proxy in Mexico and the rest of the world, has been an essential element of their political success and military survival. Information technologies and the Net have emerged as a potent weapons for guerrilla armies and governments alike. It was inevitable that the United States government—as the principal sponsor of the neo-liberal world order—would begin to sense the threat and opportunity that the Internet and the tools of the information revolution represent.

In many respects infowar is a very old idea. Propaganda, disinformation tactics, psychological warfare and signals intelligence have been essential elements in the military arsenal for centuries. The computer technology on which the information revolution is based was created in the crucible of the Allies' infowar campaign against Nazi Germany. In the name of national security, for the last five decades Western governments have funded what may be the most secret and powerful infowarriors we have ever known. One of Canada's most closely guarded secrets is the Communications Security Establishment. The organization's existence is a matter for public record, of course, and it's well-known that it employs some 900 people, is headquartered at the Sir Leonard Tilley Building in Ottawa, and that it's directly responsible to the Privy Council and the Minister of Defence... but that's about it.

CSE is one of those super-secret spy agencies that Western governments set up during the Cold War to counter the Soviet threat that never came. It was created in an age of rampant paranoia, and is one of the last holdovers of that great fear. While it's known to be involved in signals intelligence—the gathering of information from radio transmissions and telephone and computer networks—its exact mandate remains a secret. Like the National Security Agency in the United States, CSE is supposed to spy exclusively on foreign signals, but because their mandates are secret, there's no way to know for sure how they define 'foreign.' Are signals that originate overseas, but terminate in the U.S. and Canada—signals like electronic mail and Internet feeds—considered fair game by the techno-spooks? In 1973 alone the NSA retrieved more than 23 million individual communications, and this doesn't represent the total number of messages represented, but those *retained for further study*—and all this twenty years ago, when the volume of electronic communications were a fraction of what they are today![10] With a budget reportedly greater than that of the Central Intelligence Agency's, the NSA is the biggest and most powerful intelligence organization in the world. With a digital drift-net stretching around the globe, it is inevitable that almost everyone who participates in the information economy has, at some time, been caught up in America's defensive infowar. The new military realities of the information age extend far beyond passive communications monitoring.

Infowar strategies are predicated on two main assumptions. The first is that the new information technologies can amplify the military effectiveness of small groups and countries, making them a potential threat to national security out of proportion with their actual size. To some extent this is the main thrust of Swett's *Strategic Assessment*, and the uncomfortable reality—to North American corporate interests—represented by the Zapatistas. The Net, as the infrastructure of the global economy and neo-liberal economic expansion is thus, by default, the inevitable theatre of international insurgency. The second assumption is that advanced information societies are particularly dependent on high-technology, telecommunications and information, making them vulnerable to infowar offensives. Information warfare is aggressive action with the goal of achieving information superiority "by affecting adversary information, information based processes, information systems and computer based networks while defending one's own information, information based processes, information systems, and computer based networks."[11] It could involve hacking, disinformation, the planting of viruses in sensitive systems or just bombing computer installations into dust. The arsenal includes electromagnetic pulse bombs that can destroy an enemy's communications infrastructure while leaving structures intact, directed pulse weapons to disable armoured vehicle firing systems and computerized navigation and avionics on the battlefield.

Swett's report was a powerful exhortation for the military establishment to get with the program. Though many of the tools of the information revolution, and the Internet itself, are the direct progeny of military research, his observation that "the Internet could also be used offensively as an additional medium in psychological operations campaigns and to help achieve unconventional warfare objectives" was a novel idea.[12] The global reach of the Internet gives military infowarriors the perfect avenue into sensitive information systems, and every hacker knows—and more than a few have proven—that no firewall or data security system is infallible. In addition to its value as a vehicle for psychological warfare and propaganda, the Internet can also be used as a delivery system for viruses, 'logic bombs,' and direct sabotage. The infowarrior, therefore, has more in common with malicious teenaged hackers than gun-toting grunts in the field.

Of course it's equally true that American information systems are just as vulnerable as anyone else's. The level of technological development in the United States seems to make it a particularly vulnerable target for infowar attack. Information technologies have the tendency to amplify the potential of individual operators or small countries to cause havoc, and all roads on the Net inevitably lead to the United States.

> These systems, especially the military's systems, are under constant attack and are vulnerable to penetration, manipulation, or even destruction by determined hackers. In the future, these may not be unstructured anarchic amateurs but well-paid network-ninjas inserting the latest French, Iranian, or Chinese virus into Compuserve or the Internet.[13]

While such a large-scale, destructive attack is still the stuff of speculation and science fiction, American information system security has already been compromised in the past. In one of the most celebrated cases in 1986—the one that made author Clifford Stoll a literary star—a small group of German hackers in the pay of Soviet intelligence foiled some of the most secure networks in the United States, including NASA. The hackers were eventually apprehended, but not before they conclusively demonstrated how insecure computer networks could be. A decade later, the American military is even more reliant on information technologies, the weapons of infowar can be purchased at any electronics store around the world, and, on the Internet, America is just a server away.

CHAPTER FIVE

Virtual Jackboots

I. ZOG Day Afternoon

IT WAS 1994, and a virulent infection seemed to be spreading throughout the Internet. The echo of virtual jackboots reverberated throughout the on-line community, and in newsgroups like alt.revisionism and alt.skinheads, one thing had become clear—the Nazis had arrived. Long-time users who had always considered the Net's comfortable congeniality to be a given recoiled in shock and horror as white supremacists and neo-Nazis posted propaganda and sought new recruits... but they shouldn't have been surprised.

North American ultra-rightists, particularly white supremacists, had been using computers and electronic bulletin board services to quietly circulate hate literature and correspondence throughout the United States and across the border into Canada for more than a decade. The Aryan Nations Liberty Net, established by former Texas Ku Klux Klan Grand Dragon Louis Beam in the early 1980s, actually predates PeaceNet and the Alliance for Progressive Communications. Indeed, it may have been the first political computer network in North America. Canadian neo-Nazis were routinely using computers to bypass border controls on the importation of hate propaganda by 1984, and well-established BBS networks such as Fidonet and WWIVnet had become popular forums for crypto-militiamen and virulent white supremacists alike.

The Net has always been a reflection of off-line society, and there are no ideas on-line that aren't expressed in society as a whole. However, the same tendency of the Internet—and, to a lesser degree, BBS networks—to enfranchise activists, progressive organizations and women, has raised the voices of the ultra-right into the public consciousness. The voices were already there, but thanks to the Net's decentralized discourse, they could be heard at equal volume to everyone else's. So it was only a matter of time before this uglier reality of life off-line began to intrude on the civil environs of the Internet, and while white supremacists had been coming on-line in ones and twos for some time, the full chorus of their screed only became obvious in 1994.

A series of incidents on the National Capital Freenet had the Ottawa-based public access provider's administration scrambling to find a response to the growing presence of racists on their system. At the centre of the controversy was a 22-year-old skinhead and self-proclaimed racist named Jason Smith, a member of a Canadian offshoot of the notoriously violent Detroit-based Northern Hammerskins. Smith began posting messages on the can.politics and alt.skinheads newsgroups attacking minorities and gays, and seeking recruits for his group and for the Heritage Front, Canada's leading neo-Nazi organization.

His posts were a provocative call to action, exhorting all white Canadians to rise up against their Jewish and non-white "oppressors" in the name of white power. "It is time we stopped our apathetical [sic] views of ethnic diversity and assimilation and did something about it," he wrote, "Stand up for yourselves, be proud of who you are, and your accomplishments. It is because of our very compassion and gullibility that we have come to this point. We are constantly trying to present a good face to the world, trying to be the 'Good Samaritans,' all at the expense of ourselves. Perhaps it's time we were selfish?" In another post to alt.revisionism, he made the object of his hate clear: "The Jew is a parasite that must be detached from its host in order to heal the disease it has brought upon the people."

The Freenet was in a bind. On one hand, it existed for the express purpose of guaranteeing Internet access to anyone who might want or need it. However, the hate-filled diatribe and personal abuse spewing from Smith's keyboard was more than it could ignore. The matter was ultimately referred to the Ontario Human Rights Commission, with no result, and though Smith soon abruptly disappeared from newsgroups he had haunted for almost a year, it was never resolved. Smith's National Capital Freenet account is still active.

Though the medium is new, the message is very old. Always careful not to cross the rhetorical line between advocating racism and inciting racial violence—a distinction of great importance in Canadian law—Smith was nevertheless the ideological descendant of a political tradition stretching back beyond the founding of the Ku Klux Klan in 1865. Organized racism and fascism emerged in the 1980s as a rare and frightening social threat. Gone were the vaguely comical Nazi-wannabe costumes and pseudo-serious posturings of George Lincoln Rockwell's American Nazi Party, replaced by a violent militancy wearing Doc Martens boots and bearing automatic weapons and explosives. Violence, whether real or rhetorical, is a central element in the ultra-right's discourse which looms like a thundercloud over the Net.

It would be misleading to suggest that the ultra-right is a single, monolithic ideological entity. It represents a broad spectrum of ideas and opinions, from the radical libertarian anti-government stand of the so-called 'Patriots' to the violent racist terrorism of the Silent Brotherhood. It is not always *explicitly* racist or anti-Semitic, as in the case of the Montana Freemen, but undisguised bigotry is often at the centre of its theory and program. Nevertheless, these differences are little more than variations in emphasis—scratch a militiaman and you'll find a Nazi just under the surface.

The basic tenets of the movement are an unwavering opposition to government authority, a profound conviction that the 'white nations' of North America and Europe are being held hostage by an international — usually Jewish—conspiracy of bankers, communists and internationalists, and a fundamental intolerance for anyone who does not conform to its ideals of ethnic and social purity. Central to the ultra-right's belief system is the conviction that the American (and Canadian) government is a puppet of the Jewish conspiracy they call the 'Zionist Occupation Government.' They believe ZOG is behind all of the social ills facing the world and is bent of the destruction of the Aryan culture and the 'mongrelization' of white racial purity.

Clutching William Pierce's novel *The Turner Diaries* close to its heart, the movement's goal is to purge America—and Canada and Europe—of social and racial corruption. In Pierce's fiction, a fanatical band of white supremacists called The Order 'cleanses' North America of Jews, blacks, minorities and immigrants. Cities are devastated in thermonuclear strikes, and in Los Angeles on the 'Day of the Rope' the garroted bodies of 'race traitors' are strung from trees like Christmas ornaments. *The Turner Diaries* is fiction, but for this generation of ultra-rightists it is more... far more. For them, the book is a manual for race war, the inspiration for their militancy, a bible. The Silent Brotherhood, the terrorist gang that gunned down Denver talk-show host Alan Berg in front of his home in 1984, consciously modelled itself on the fictional Order. Timothy McVeigh, the convicted bomber of the 1994 Oklahoma City Federal Building, also read every page of the *Diaries* like revealed truth.

At the heart of the ultra-right discourse are 'the 14 Words,' the movement's rallying cry. Coined by David Lane, a Silent Brotherhood terrorist serving a 50-year sentence for conspiracy, racketeering and his part in the murder of Alan Berg, the slogan carries the weight of a mystical incantation: "We must secure the existence of our people and a future for white children." Smith's messages, sometimes 'exposing' an alleged international Jewish conspiracy, with veiled exhortations to violence against gays, immigrants and minorities, waved the banner of white power and rarely departed from the 14 Words. In terms of form, style and content, they were indistinguishable from the pamphlets and flyers frequently handed out by white supremacist militants at county fairs and on college campuses. However, Smith represented something different, something frightening in its novelty—a young, educated neo-Nazi with the means and the motivation to use the new technologies as a weapon in the war against non-whites. "The older generation who started the [white supremacist] movement is taking a back seat to the younger generation," Smith said at the time. "I realized the potential of the Internet both for its networking capabilities and for its information distribution potential. The Net provides an audience that I can't ignore... it's a good tool for propaganda."

This was something new—a white supremacist militant staking a claim to the Net. Smith was a power on the alt.skinheads newsgroup, pulling conversations into line, and intimidating those anti-racists who stumbled across it. The forum covered the whole breadth of skinhead culture, from passionate discussions on bands, information on where to buy hard-to-get recordings and fashion. One discussion, that ran for several months, dealt with the important question of whether skinhead girls should shave their pubic hair — the general consensus was 'yes,' but only one skinhead girl participated in the debate. At the time when Smith was the newsgroup's de facto leader and moderator (being an 'alt.' newsgroup, it is not officially moderated), most postings were not *explicitly* racist, but he ensured that white power and neo-Nazi campaigning were never far from the surface. The goal of Smith and his colleagues was to mix hate with a palatable blend of high-technology, rebellion, rock and roll and a sense of community. "It's a new approach. It has a lot to do with our generation," said Milton Kleim, then the neo-Nazi National Alliance's most prominent on-line propagandist. "The main factor in this move toward new technologies is a certain elite which is filtering up as social conditions in

America degenerate and as Generation X individuals like myself and Jason move into the mainstream."

Smith and Kleim were far from mainstream, but they were part—perhaps a big part—of a transformation that had begun to sweep through white supremacist ranks in the mid-1980s. The structure of the old movement was defined, if anything, by factionalism and disunity. Each Ku Klux Klan klavern (or local chapter) across North America operated more or less independently, and few of the putative national organizations, like the National Socialist White People's Party or the National Alliance, had more than a few scattered members outside of their head offices. With the recruitment of disaffected youth—often skinheads—and the emergence of the Net as a mass medium, that began to change. The white supremacist movement on the Net is both more intellectually sophisticated and self-conscious than ever before. "The movement has been very fragmented in the past, and to be truthful, it hasn't always been a brains trust. But we are becoming more organized as time goes by," Smith observed. "What happened was a lot of the older generation in their 50s and 60s who started this way back when are beginning to take a back seat to the younger generation. We're generation X, we're the most educated generation that has come along yet, and we see the errors that our forerunners made. We were educated in the same educational system as everyone else in this country, and so we know the perceptions and the way people... I was subjected to those same anti-racist and anti-fascist programs, those rights for gays programs, and the multicultural programs which try and get people to accept these things. We know exactly what the opposition is trying to do, because we were brought up with them."

This new generation of ultra-rightist militant has sought to use the Internet to put Louis Beam's doctrine of 'leaderless resistance' into practice. This doctrine holds that an anti-government force needs to build flexibility and stealth into its very organizational structure—by having no formal structure.

> A system of organization that is based upon the cell organization but does not have any central control or direction, that is in fact almost identical to the methods used by the Committees of Correspondence during the American Revolution. Utilizing the Leaderless Resistance concept, all

> individuals and groups operate independently of each other, and never report to a central headquarters or single leader for direction or instruction, as would those who belong to a typical pyramid organization.[1]

While on-line neo-Nazis typically share a phenomenal level of ideological consistency—most, if not all, believe in the '14 Words'—there is no central organization. Smith, and to a lesser extent Kleim, operated as a freelance racist militant despite his affiliation with the Northern Hammerskins, whose whole Canadian membership doubtless numbered considerably less than a dozen. However, for a movement committed to leaderless resistance, the Net was a godsend. A movement consisting of any number of small cells and regional organizations at the fringes of society, would ordinarily have a tough time creating any cohesion.

Since the founding of the Klan at a Christmas party in Pulaski, Tennessee in 1865, the ultra-right has sought to create a sense of solidarity, community and brotherhood through propaganda and ritual. At the centre of the white supremacist's political experience, however, is the all-important rally and cross-burning, a ritual opportunity for neo-Nazis to reaffirm their commitment to the movement, collect dues, sell souvenirs and memorabilia and frighten the wits out of their race enemies. Rallies, like the annual Aryan Fest at the Aryan Nations' compound in Hayden Lake, Idaho, or the notorious gathering in Caroline, Alberta, in September 1990, and skinhead rock concerts and parties are, above all, social events. Though they often draw handfuls of participants from distant locations, they are above all local—at best regional—events. With that in mind, rallies, as frightening as they may be to non-racists, are inadequate as means of mobilizing mass support for the 14 Words. For the neo-Nazis of Smith's generation, and the established organizations that want to reach out to them, the Internet offers an opportunity to break out of the movement's historical parochialism.

The Internet is cheap and easily accessible. Anything that can be printed can be placed on a web site, e-mailed to colleagues and fellow travellers, or posted on Usenet newsgroups—and at extremely low cost. The same economies of scale that prevented individual political activists and organizations across the political spectrum from reaching a broad audience using traditional media guaranteed in the past that there were relatively few neo-Nazi publications in circulation. However, the information revolution has liberated activists of all stripes from these media, and just as the Net has encouraged a flowering of dissent and opened up the discourse of the left, it has stimulated the metastasization of the ultra-right discourse.

The ultra-right's old guard has only grudgingly embraced the Net. The National Alliance in the U.S. and the Heritage Front in Canada have established Internet presences—though the latter are generally promoted by American proxies, due to relatively stringent Canadian laws against hate propaganda. Indeed, the Net has permitted Canadian and German racists to publish their opinions beyond the reach of their respective countries' laws, while still reaching an audience at home. The National Socialist White People's Party (formerly the American Nazi Party), the National Alliance, Heritage Front and others publish listserv newsletters and book publishers and skinhead record companies offer the literature and music of hate to anyone with a credit card. Yet the vast majority of ultra-right militants on the Internet are freelance propagandists like Smith, operating on their own initiative and pursuing a campaign of 'leaderless resistance.' If they even belong to established organizations, their on-line activities are neither controlled nor explicitly sanctioned by the traditional leadership.

II. The Big Lie

If neo-Nazis possessed a sense of irony, they might get a chuckle out of their unanimous insistence that the Holocaust, in which six million Jews, and millions of Gypsies, homosexuals, Slavs and political prisoners were murdered by Nazi Germany, is a hoax. While they venerate Adolf Hitler for the Final Solution, —they celebrate the dictator's birthday every year and Kleim once called his ascent to power the most important event in modern history—they deny that the Holocaust ever occurred. While it isn't an explicitly anti-Semitic discourse, and many self-styled 'revisionists' insist that they have nothing against Jews, Holocaust denial is, in fact, the respectable face of the ultra-right. It's neo-Nazism in professorial tweeds, hate propaganda in the raiment of scholarly inquiry... and on the Net, it's astonishingly common.

It's hard to say how much of the general population doubts the historical veracity of the Holocaust. A survey by the Roper organization for the American

Jewish Committee found that in 1992 twenty-two percent of American adults believed that it is "possible that the Nazi extermination of the Jews never happened."[2] The Roper survey has since been conclusively shown to be inaccurate—many experts have doubts about both its methodology and the question's wording—but the *grey* area it represents is an indication of the ambivalence that many non-Jews feel about the Holocaust. Many who accept that the event occurred are nonetheless doubtful of the historically accepted figures for how many people actually perished in the death camps. Indeed, for some people the vast horror of the Nazi crime is almost incomprehensible. For the ultra-right, such doubts are an opportunity to cultivate hate and fear, fertile ground for the seeds of neo-Nazi anti-Semitic propaganda.

"I'm not a racist, but that should be beside the point," Greg Raven said, more than a little disingenuously. He is the president of the Institute for Historical Review, a pseudo-academic organization dedicated to rewriting history to suit its ideological tastes. Despite its impressive name—doubtless intended to *sound* like the legitimate Institute for Historical Research—it has no credentials, no affiliations with academic institutions, no legitimacy. It is a sham. Raven insists that the IHR is not solely interested in denying the Holocaust, though the vast majority of articles in the Journal of Historical Review, the IHR's principal organ, seek to deny it. Those few articles that are not specifically related to Holocaust denial are usually general anti-Semitic tracts like "Zionism's Violent Legacy" and "The 'Jewish Question' in 15th and 16th Century Spain." "Objectively, [the Holocaust] is one of the most important issues facing us today," Raven argues. "An endless parade of World War Two and Holocaust-based issues is washed across average people. It's used in foreign-policy decisions and policy in Israel. We're constantly being urged to learn the lessons of the Holocaust, but we have to know the truth of it. If the Holocaust was given what we feel has its proper emphasis as a subset of World War Two, then it can't be used for political gain... let's find out what *really* happened."

What really happened, according to Raven and other deniers, is that the Holocaust — one of the most completely documented events in human history —never occurred, that it is a myth cooked up by Zionists to win international sympathy for the world's Jews. Deniers typically maintain that the evidence of extermination was faked and that the remains of gas chambers and crematoria are not what they seem. Some go so far as to argue that the death camps of Auschwitz, Sobibor and Belsen were really resorts and settlements where happy Jews frolicked in heated swimming pools and watched vintage movies while the rest of Europe was consumed by war. What the deniers fail to explain, of course, is just what happened to the eleven million Holocaust victims who simply vanished from the face of the earth.

They don't have to, since it is doubt and not knowledge that they attempt to promote. Despite Raven's protests, Holocaust denial is plainly and explicitly an element in the ultra-right's anti-Semitic agenda. In order for the Holocaust to be a hoax, it must be perpetrated by a vast international conspiracy whose interests are served by such a mass deception—an international *Jewish* conspiracy. The mission of the IHR is to make denial, and thus anti-Semitism, palatable. As Harold Covington, leader of the National Socialist White People's Party wrote, "the real purpose of Holocaust revisionism is to make National Socialism an acceptable political option again."[3] Holocaust revisionism is simply neo-Nazi propaganda without the violent rhetoric and crass slurs, it is the pleasant face of hate that the ultra-right hopes will somehow make racism and genocide "an acceptable political option."

Raven's insistence that the IHR is not motivated by racism is particularly absurd in light of the organization's history. Founded in 1979 by Lewis Brandon, better known as British neo-fascist William David McCalden, the institute was long associated with Willis Carto, one of the American ultra-right's most prominent leaders. In his quest for an appearance of legitimacy, Raven has sought to distance the IHR from Carto since the latter was ousted from a leadership position in the organization in 1994. Raven weakly insists that Carto "never had much to do with editing or the content of the journal, and he wasn't really the founder." However, IHR letterhead proudly named Carto as founder for several years. Despite Raven's desire to rewrite his *own* history, the institute's principal function on the Internet is to serve as a clearinghouse for the kind of ideas long promoted by Carto. Indeed, the IHR maintains a close association with the Noontide Press, which publishes such neo-Nazi classics as *The Protocols of the Learned Elders of Zion*, *Our Nordic Race* and *Hitler: The Unknown Artist*. Carto's wife Elisabeth served as the Press' treasurer and business manager.

The *Journal of Historical Review*'s editorial board and staff is a who's who of the pseudo-intellectual ultra-right, and includes such dubious luminaries as the French former academic Robert Faurrison and Sweden's Ditlieb Felderer, both convicted of hate-related offenses in their respective countries. Raven himself has often seemed quick to defend Hitler, whom he evidently believes to be a misunderstood genius, maligned and slandered by the international Jewish conspiracy.

> I will say, however, that he was a great man... Certainly greater than Churchill and FDR put together, and possibly the greatest leader of our century, if not longer. This is not to say that he was perfect, but he was about the best thing that could have happened to Germany.[4]

There can be no doubt of either the IHR's ideological agenda, or of the role played by the Internet. "Anyone who wants to get a important message out has to consider using the Net," Raven says. "We publish a small journal for a select audience, but thanks to the Internet we can reach a far bigger audience than we could in print."

The Canadian Holocaust denier and anti-Semitic propagandist Ernst Zundel has certainly taken that observation to heart. After years of high-profile legal battles, including a subsequently overturned conviction under Canadian anti-hate laws in 1985, Zundel has turned to the Net as the principal means of getting his message out. Like Raven, the portly, clownish Zundel is neither a writer nor a particularly deep thinker. He is simply a mouthpiece, a conduit for anti-Semitic propaganda. While most of Zundel's legal problems have been the result of his publishing of hate propaganda that may violate Canadian law—notwithstanding a 1992 Canadian Supreme Court decision that set aside the law that making the publication of "false news" a crime—he has found that these can be easily circumvented on the Internet. The "Zundelsite," a Web site devoted to the promotion of Zundel and his ideas, is actually hosted on a server in the United States, placing it conveniently outside the reach of Canadian law.

The Zundelsite provides links to denial and ultra-right-related sites around the world in both English and German. The bilingual site represents the dichotomies that lie at the heart of the deniers' discourse in particular, and the ultra-right in general. Zundel and his supporters insist that the denial movement is international in scope, but the site is almost overwhelmingly devoted to his continuing legal difficulties in Canada. Zundel is paranoid... paranoid of the courts, paranoid of the press, paranoid of Jews. He rejects the suggestion that he is an anti-Semite, preferring to portray himself as a guardian of truth, persecuted by a conspiracy of Jews; yet the Zundelsite logo—a black 'Z' in a white circle on a red field—is clearly intended to resemble the Nazi swastika flag, and his paranoia takes the form of a firm belief in an international Jewish media conspiracy. When he refuses to talk to the press, Zundel makes a point of singling out those people he considers to be his enemies: "My policy now is that when a Jew calls for an interview, I say good-bye and hang up the phone."[5]

III. Net Nazi Number One

For three years, a clean-cut university student from central Minnesota named Milton John Kleim Jr. was the most energetic and, at times, the most articulate on-line propagandist the ultra-right had ever seen. Kleim's brief career embodies the whole recent history of the neo-Nazis' Internet campaign. Canadian journalist Crawford Kilian called him "Net Nazi Number One," and the nickname stuck. Kleim wore it like a badge of honour, flaunting his intelligence—rare in a hatemonger—and youthful erudition in the service of hate. In 1995, the Anti-Defamation League of B'nai Brith called him an "innovative propagandist," observing that he

> thinks he has a method that might work. Kleim, a recent graduate of St. Cloud State University in Minnesota who likes to write earnest-sounding tomes that read like class papers, tries to project an image of a rational, scrubbed and shined, eager young activist... Kleim floods the newsgroups with messages attacking Jews, "Jewsmedia" and non-whites, openly calling for authoritarian government...[6]

As much as Jason Smith seemed to hold out a promise to the ultra-right of a new age of on-line militancy that could reach millions of Internet users across borders and around the world, Kleim appeared to be its fulfillment. He wrote and posted with extraordinary energy, willfully picking fights with the growing ranks of on-line activists. As far as he was concerned, the Net was an opportunity that he simply couldn't pass up. The possibilities were almost endless, and Kleim soon found that he could reach, and he hoped, hector thousands of people at minimal cost. He had discovered what the activists on the left knew; the Net could circumvent traditional media and serve as a powerful medium for disenfranchised voices.

Kleim's brief career as the ultra-right's leading Internet propagandist encompasses the recent history of on-line white supremacist agitation to date. Net Nazi Number One literally wrote the book on on-line neo-Nazi activism. His 1995 article "On Tactics and Strategy for Usenet," posted to several Usenet newsgroups, and archived on many ultra-rightist FTP and Web sites, spells out exactly how he believed the Internet race war should be waged.

> USENET offers enormous opportunity for the Aryan resistance to disseminate our message to the unaware and the ignorant. It is the only relatively uncensored (so far) mass medium which we have available. The State cannot yet stop us from "advertising" our ideas and organizations on USENET, but I can assure you, this will not always be the case. NOW is the time to grasp the WEAPON which is the INTERNET, and wield it skillfully and wisely while you may do so freely.[7]

It is a powerful article, exhorting neo-Nazis to practice the on-line equivalent of guerrilla warfare, making quick hits in a variety of newsgroups, but avoiding contact with the 'enemy' (anti-racist activists) wherever possible. Kleim lays out a strategy for first contact to recruit members, telling neo-Nazis to e-mail users who post sympathetic articles, welcome them to the fold and provide recruiting information. While warning his comrades not to cross-post articles to irrelevant newsgroups lest they earn the contempt of other users, Kleim hammers home his central point that propaganda must be kept simply and posted frequently. "SUSTAINED, electronic 'guerrilla warfare,' 'hit and run' style, using short, 'self-contained' posts is a major component in our struggle."

Following his own instructions, Kleim sustained his guerrilla assault on Usenet almost every day for more than two years. He established an Aryan News Agency, and worked on the National Alliance's *Turner Diaries* Web site, but Net Nazi Number One always felt most comfortable in the raw text environment of Usenet. "I do see Usenet as being divided into three camps—the White Nationalists, the so-called anti-racists, and the people who are sort of in the middle, and sometimes they go either way," he observed. "It's sort of like a bar with thousands of rooms, and you can always find somewhere to brawl with someone. I'd say a good fifty to sixty percent at least is just somebody wanting to argue, and they just start something. That's probably the main reason why so many flame wars get going on Usenet. Then, of course, there are some people who really do have a beef with us."

However, Kleim's major contribution to the neo-Nazi cause were his tracts and essays, posted to Usenet and reproduced on ultra-rightist FTP and Web sites all over the Net. They were often self-consciously coy, dancing around the issues of hate and racism to make them somehow attractive to potential recruits in search of a political identity and racial community. His "National Socialism Primer," written in a style more reminiscent of a summer camp marketing brochure than a hate group's political tract, remains one of the movement's most popular—and revealing—recruitment documents. Kleim goes to great lengths to show that "National Socialism is based on love of one's own kind and the Creator's benevolent natural Order, *not* hatred," but never quite leaves the central racist program of the neo-Nazis behind:

> Jews, members of a people who have chosen themselves to rule the World, comprise a majority of the World Manipulators, with substantial numbers of Aryans and Asians intoxicated with Judaic thought participating in this obscene racket.[8]

Kleim always insisted that he wasn't a Nazi. "[The term] represents something I'm not," he once said. "I'm not a loony kook walking around in a uniform and screaming 'gas the kikes.' I don't do that. I don't believe that. You know, I seek a much different worldview than that. National Socialism is exactly what it says—nationalistic socialism." For all of his protests and intelligence, however, Kleim's on-line activism never departed from the ultra-right's message of hate.

Born in 1971 to a conservative German family in Sacramento, California, Kleim is typical of the current generation of neo-Nazis, both on-line and off. His initiation into the ultra-right was less a question of firm belief than a search for identity. Today, Kleim says that at seventeen he was looking for something that he couldn't quite express and restlessly moved from one political cause to another—including work with the Democratic party on the 1988 Dukakis-Bentsen presidential campaign. Andrew Oakley's book *88,* casually picked up by the alienated white teenager in central California, contained a comprehensive listing of American hate groups that offered something that other causes seemed to lack—identity, a cause, a movement.

The ultra-right's recruits tend to be young, usually middle class and male. There *are* many young women in the movement and organizations like Women for Aryan Unity that promote white supremacist values for women, but there's something in the militaristic, violent Nazi iconography that appeals to young men. Indeed, the ultra-right's principal selling point is not racism — the uglier side of the discourse is often played down in recruitment materials like the "National Socialism Primer"—but identification with a mythopoeic teutonic warrior ideal. Just as paleo-Nazi recruitment propaganda emphasized the visceral power of a mock-Nietszcheian superman, the neo-Nazis promise empowerment. By joining, young men become Aryan warriors in a divine mission to deliver their people from the grip of ZOG. Rather than remaining anonymous participants in a bourgeois economy divested of the last shreds of the American Dream, the ultra-right promises that they will be special... they will be *heroes.*

The ultra-right has always grown at times of great social dislocation and economic crisis. The heyday of the Ku Klux Klan came during the Great Depression, when the optimistic expansion that had propelled America since the Gilded Age of the late nineteenth century seemed to contract in the midwestern dustbowl. The neo-fascist National Front was propelled to prominence during the desperate days of Margaret Thatcher's Britain, when the slums of Brixton and Tottenham momentarily overwhelmed London and it became abundantly clear that the sun had finally set on the British Empire. The American ultra-right peaked during the mid-1980s, at a time when Lyndon Johnson's Great Society was systematically dismantled in the raw utilitarian atmosphere of fiscal conservatism and Reaganomics. Young people like Kleim do not become neo-Nazis out of hate. They see the promises of civil society and the American Dream being broken all around them, they experience the bleak ebb of a future that had been all but assured and with it the loss of economic identity. The ultra-right offers a belief system that tells them who to blame and promises an exalted status in the context of apparent decadence around them.

Kleim's search led him to Christian Identity movement, the pseudo-religious foundation of Richard Girnt Butler's Aryan Nations. Identity preaches that white Europeans—Aryans—are the true descendants of the biblical Israelites, God's chosen people. The Jews are literally the spawn of Satan, corrupting the world and conspiring to thwart the divine mission of God's people at every turn. All other non-Aryans are 'mud people,' pre-adamic and thus sub-human. Christian Identity offered Kleim just that — an identity—without directly associating him with Nazis. Identity Christianity seemed safe enough, but by 1991, Kleim had drifted into explicitly neo-Nazi groups.

In fairly short order, Kleim became acquainted with both the Internet and Pierce's *Turner Diaries*. "That did it," he said. "Ninety-five percent of the people who read *The Turner Diaries* are going to be utterly repulsed by it; for some reason the other five percent are just 'Wow!'" Kleim was one of the five percent. In 1993, he wrote his first article for the National Socialist magazine *Instauration*, and began studying Anthropology at St. Cloud State University, where one of his professors encouraged him to take his first steps on the Internet. By 1995, he was an active member of Pierce's National Alliance and generally regarded as one of the most important and influential white supremacists on the Net.

However, in the summer of 1996, Kleim shocked his erstwhile neo-Nazi comrades when he suddenly and dramatically broke with the movement. In a series of open letters posted on Usenet to his former colleagues, Kleim lashed out at the leadership of the neo-Nazi movement. "Your lives mean more to me than earning yourselves prison for Pierce or a bullet for [Canadian skinhead leader George] Burdi," he wrote. "To me, you are young Aryans with a future; to them, you are an instrument in a personal political quest. Expendable. Replaceable. You are pawns in a game played by small men with large egos. You are tools for the fulfillment of power trips, grandiose schemes, lunatic, unrealizable designs on America and the world."

Perhaps reflecting his own search for identity as an alienated youth in Sacramento, Kleim condemned the adolescent motivations of the ultra-rights' fresh recruits. "The 'movement's' entertaining aspects serve as a diversion and a release for the sexual tension all racially-conscious Aryan males experience each and every day multi-culturalism is encountered," he wrote. "In a short-sighted yet gratifying manner, cussing 'niggers' and bitching about 'the ZOG,' listening to the latest pseudo-revolutionary idiocy the drunkards have screeched out, or reading the latest moronic missive by a self-serving, money-grubbing guru are imagined to be 'fighting for the Race.' Reality, however, is that it's a cop-out, a feel-good recipe to soothe hyperactive hormones."

By autumn, Kleim had not only denounced the movement, but had renounced and repudiated the entire corpus of his propaganda. He still had some way to go, and his vocabulary was peppered with the neo-Nazi references that had been so much a part of his language for so long, yet the change was there. "I have repudiated and denounced hatred," he now says. "I'm equivocal on the aspect of race and racism, because I believe that probably 95 percent of the people in our society in our society are racist in some way—it's a matter of degree. I guess I'm still a racist in that I have preferences for certain types of people… I'm not trying to soothe my soul, but in a way I was kind of a propagandist for hire. I got caught up in it and I was doing what I was supposed to be doing."

Today Kleim calls his erswhile anti-Semitism irrational. Yet, for most of his adult life, he was an active and virulent anti-Semite, somehow justifying his hatred of Jews-as-a-group with his freely admitted respect for individual Jews. "I don't know where it came from, but I guess it really indicates how pernicious the propaganda is," he says. "It took a while to be indoctrinated, but if someone as rational as I am normally can be taken in by that stuff, what happens when someone who is average is reading it and they don't critically analyze it."

Few in the ultra-right saw it coming, but Kleim said his break with the movement and with his Net Nazi past was only part of a process of reevaluation that had begun months before. Indeed, in the autumn of 1996, it had begun to seem that the on-line ultra-right was in complete disarray, and Kleim's departure was simply a reflection of its inability to adapt to the new technologies and to its impotence as a movement.

IV. Paper Brownshirts

The ultra-right's slick, elaborate Web sites give the impression of a well-organized, mass movement at the cutting-edge of technology. They're supposed to. Impressive Web pages and heavy participation in Usenet might seem to indicate that the ultra-right is healthy and vigourous, but appearances can be deceiving. The most energetic on-line propagandists like Kleim are usually independent freelancers, who may be associated with the neo-Nazi movement, but are not directed by it. Moreover, it's almost impossible to know exactly how many people are behind a Web site, whether it is the work of an organization, or an individual. On Usenet, one person with a handful of different Internet accounts can create the impression of a vigourous movement by having a conversation with himself. The technique—known as pseudo-spoofing—is a favourite neo-Nazi tactic. Even one user, using a single access account, can throw a disproportionate weight around the net, by *spamming*, or cross-posting to every newsgroup on Usenet.

The neo-Nazis themselves can't say if their on-line propaganda and recruiting efforts have had any real effect for the white supremacist movement as a whole. Smith said he had received some encouraging e-mail, but ultimately conceded that he spent most of the time preaching to the converted while those who disagreed with his opinions ignore him. In 1994, Kleim blustered "If I can cause one person to open his eyes and just question what the establishment issues, then I have succeeded and my efforts are justified." Today he concedes that he probably wasn't responsible for recruiting more than ten people for the ultra-right, and even they were just people who asked for additional information.

There probably has never been a time when the racist ultra-right was a single, unified movement, and today, despite well documented ties between the Klan, Holocaust deniers, skinheads and white nationalist organizations like the Aryan Nations and the National Alliance, the ultra-right is as fractious and muddled as it has ever been. The expulsion of Willis Carto—reportedly at gunpoint—from the IHR is merely symptomatic of the inability of racists to maintain a cohesive and effective organization. Most ultra-right groups are ephemeral organizations generated by the socio-economic conditions of the time. Their lack of a consistent program and ideological centre beyond the immediate appeal of hatred for the 'other' makes them both volatile and fragile. Canada's Heritage Front, for example, loomed into the media spotlight in the mid-1980s as the reinvigorated threat of fascist direct action. After reaching its peak of effectiveness around 1988, it was rocked with defections and scandals—notably the revelation that one of its leaders had been an informant for the Canadian Security and Intelligence Service—that brought the organization to its knees. Today, despite the appearance of a Web site and an electronic mailing list, the Heritage Front is little more than an extremist group on paper and in the minds of its leaders.

The ultra-right's activities on the Internet are at best haphazard. The neo-Nazi movement on the Internet is an iceberg inverted: apparently solid and threatening on the surface, but without substance or support below. Despite the best efforts of its few noisy on-line propagandists, who never number much more than a dozen on Usenet and less than two hundred on the Web at any one time, the ultra-right's leadership has shown remarkably little interest in the new technologies. "I was never acknowledged by the [National Alliance's] national office or given resources," Kleim recalled. "I was never given a dollar to do anything I did. Everything I did was either through the school or on my own funds. I did all of that on my own. I wrote 'On Tactics and Strategy for Usenet' on my own. I wrote the 'National Socialism Primer' on my own. I was never given any support from the national office. In fact, every time I talked to Pierce, I was made to feel bad. This guy was condescending, it was like I was doing something wrong and he had to correct me. Toward the end, I was trying to convince the national office to officially open a propaganda office focusing on the Internet, and it always fell on deaf ears. I don't know if it was because they didn't consider it important enough, or they considered me incompetent. There was an editorial in the National Alliance Bulletin in May [1996] that was directed at [on-line propagandists], about how there were certain people on the Internet who were creating a bad image for the movement."

On-line neo-Nazis are all about image. The principal focus of most of the *organizations'* web sites is merchandising. The National Alliance will sell you a paperback copy of *The Turner Diaries* for $88.00, and Ernst Zundel would like nothing more than to take your money in exchange for one of his videos. For most young neo-Nazis—particularly skinheads—image matters more than ideology. It's an opportunity to get tattoos, wear taboo symbols and frighten your neighbours. Indeed, in Kleim's words, most neo-Nazis are simply hobbyists, weekend Aryan warriors willing to pay through the nose for a third-rate paperback.

Nevertheless, the ultra-right's message of intolerance does find resonance in one group of eager listeners, however small. Timothy McVeigh, convicted of the April 1995 bombing of the Alfred Murraugh Building in Oklahoma City, was an enthusiastic fan of *The Turner Diaries* and has well documented links to neo-Nazi organizations. There are even reports that admirers of Robert Matthews and David Lane's white supremacist terrorism have resurrected the Silent Brotherhood. The movement itself isn't going anywhere, but there remains a lunatic fringe bent on violence and destruction. It is easy to overstate how important on-line campaigning and ultra-right ideas are to the bombers and terrorists. As was the case with such quasi-political groups as the Weather Underground in the 1960s and the Symbionese Liberation Army in the 1970s, the ultra-right fringe is probably motivated less by ideology than psychopathy and a vague sense of frustration and undirected anger.

V. Electronic Frontier Justice

At first look Ken McVay is an unlikely hero in the war against racism and intolerance. An ex-marine, former county sheriff in the American northwest, and one-time convenience store manager, he looks above all like one of the countless refugees from the 1960s who have found sylvan isolation in the rainforests of British Columbia's Vancouver Island. Graying blond hair, long and unkempt, frames a gaunt, lined face and intense eyes as hard as steel like a gunfighter in a spaghetti western. This is a man who long ago abandoned any interest in his personal appearance, casting it aside as an unnecessary complication to his pursuit of the mission. McVay doesn't suffer fools gladly, and what irritates him the most are the neo-Nazis and hate propagandists who have begun to use the Net to spread their message of intolerance. Put simply, they offend his intelligence, and they are living evidence of an ugly violence that McVay believes has no place in civilized society.

There is no consensus on what the response to the ultra-right should be. In countries like Canada and Germany which already have strict anti-hate laws on the books, lawmakers are investigating legal remedies; the Canadian branch of the Simon Wiesenthal Center—the most famous anti-Nazi organization in the world—called on that country's government to regulate Internet service providers as broadcasters, thus making them responsible for content. The application was rejected. Today, the Center's head office is careful to distance the centre from the Canadian petition. "We have never called for censorship," insisted Mark Weitzman at the Simon Wiesenthal Center

"We don't want that. But if there are laws that exist that appear to be broken, then I don't think there is anything wrong with pursuing it."

The ultra-rightists themselves gloat that, either way, there's no way to keep them from getting their message out. "If they try to censor us, the government makes us stronger because we'll have to operate in a more organized fashion," said Kleim shortly before he left the movement. "yet if they don't censor us, they give us publicity."

That's a risk McVay is willing to take. He has pursued a personal crusade to discredit neo-Nazi propaganda and Holocaust denial on the Internet since 1992. In 1995, he founded the Nizkor Project[9] to monitor neo-Nazi activities on the Net, and to provide a vast library of historical documents and information resources to confront hate propagandists on Usenet and interactive public forums. Every ultra-rightist's Usenet post is archived and catalogued, along with detailed reports on hate groups by the Anti-Defamation League and Anti-Racist Action. Volunteers around the world gather information and collaborate through a high volume listerv. The Project is predicated on one of the fundamental assumptions of the information age—that information can itself be fashioned into powerful tools or, in this case, weapons.

Other anti-racist groups have turned their attention to the Internet. The ADL and the even the Simon Wiesenthal Centre have ongoing investigations, and Hatewatch has catalogued the often Byzantine relationships between ultra-right organizations, but none have taken such aggressive, direct action, discrediting the neo-Nazis discourse and proponents at every turn. "You fight ideas with ideas," McVay says. "Sure, the kind of things the Nazis are saying are potentially dangerous, but no more so now that it's on the Net. These guys are their own worst enemy. Every time they open their mouths, something stupid comes out. Now why would you want to shut someone like that up?"

"In the strictest sense, nobody has to police the Net. That's the beauty of the medium. All we have to do is provide information and point users to Nazi home page. Anyone who does it will see that they're fruitcakes." Rather than a threat, the Internet provides an extraordinary opportunity to those people who want to battle racism and fascism both on-line and off. It is, after all, easier to kill cockroaches when they crawl out from under the fridge, and when the ultra-rightists emerge from the security of their meeting halls and rallies they *are* vulnerable. The nature of the medium forces them to provide information, to participate in the discourse of civil society rather than hiding behind their contempt for ZOG. While printed books, paper pamphlets and newspapers—the ultra-right's traditional propaganda media—permitted no interaction or confrontation, anti-racists can respond *immediately* to ultra-right Usenet postings, or use the Web to create an immediate context for links to hate-related sites. In addition to its vast libraries of information on ultra-right groups and activities, Nizkor maintains links to several hate sites, as a kind of object lesson. Hatewatch provides a brief background and introduction with links to sites as diverse as Women for Aryan Unity and the National Alliance. In both cases, the ultra-right is placed in hypercontext—the information is available and the proof is just a mouse click away.

In fact, the ultra-right may be doing us all a favour. It is instructive to see on-line hate as a gadfly—as an irritating and distorted reflection of society that doesn't let us fall into the arrogance of complacency. As long as Ernst Zundel and Greg Raven deny the Holocaust, we will be forced to remember it and try to understand it. And the extreme racists and neo-Nazis do no more than force us to confront ideas that have, to a greater or lesser extent, been expressed throughout our history. In order to heal a festering wound, you must look at it to clean it out. Racism is a social sore, a cancer that we must confront so that we can someday excise the tumor.

For its efforts, McVay and the Nizkor Project have earned the derisive scorn of their opponents. Publicly, Kleim's old comrades say they consider Nizkor to be little more than a minor nuisance. They comically call their most vocal opponent Ken McOyVey, and charge that he is the well-paid hatchet man of the Jewish community, ignoring the fact that he is not Jewish, and the project is chronically underfunded. Most often, however, they cut and run. Taking Kleim's tactical analyses into account, the typical ultra-rightist reaction is to simply disappear when confronted by a Nizkor activist, and though the project provides links to many ultra-right sites, few of the neo-Nazis return the favour. Perhaps they are afraid of having their screed seen in context. Most disturbing are the threats and hate mail that Nizkor volunteers have received since the site went on-line. Characteristically, McVay calls such threats "encouraging. It's not pleasant being threatened," he says, "but you know you've struck a nerve when it happens."

Surprisingly, McVay's reception in the Jewish community, though usually positive, has been equivocal on occasion, and sometimes downright hostile. Although the Canadian Jewish Congress asked him to speak at its 1994 plenary in Montreal, its leaders have often publicly been at odds with his methods. The Simon Wiesenthal Center's Canadian director Sol Littman went to far as to criticize McVay for pursuing frontier justice, implying that the Nizkor project is a band of digital vigilantes. Indeed, what Nizkor does *is* the intellectual equivalent of virtual street fighting, but in a way that is really no different from the struggle of the *philosophes* and rationalists against ignorance and superstition in the 18th century. The theatre has changed and, because today's ignorant lies are the underpinning of genocidal mania, the stakes are higher. Most importantly, while traditional authority figures like Littman and community organizations like the CJC dither about, befuddled by the growth and power of the on-line medium and unable to settle on a coherent plan of action, the Nizkor Project is showing results.

CHAPTER SIX

Information Counter-revolution

I. Search and Seizure

ON THE MORNING of April 13, 1995 the residents of seventeen Montreal-area homes awoke to the largest raid on electronic bulletin board systems in Canadian history. Seventy-five Royal Canadian Mounted Police officers executed search warrants on the sysops (system operators) and co-sysops of eleven BBSes suspected of trafficking in pirated copyright software. No one was arrested, but fifteen people had their computer equipment and other items seized as evidence, and all were later charged.

The atmosphere on local BBSes was a strange combination of relief and horror. On one hand, sysops and users alike were thankful that they weren't caught up in the police dragnet. On the other hand, it had suddenly become clear that the police were watching. The Mounties didn't target the BBSes indiscriminantly. The raid was the culmination of an operation which had begun the previous November, when the RCMP's copyright investigation unit began calling up some of the larger BBSes. One board led to another, and before long, the Mounties were ready to reel the suspected pirates in.

According to the RCMP, the raid recovered millions of dollars' worth of pirated software and hundreds of thousands of dollars worth of computer hardware which was to be held for evidence. In fact the RCMP's estimate of the scale of the piracy operations, with some sysops allegedly receiving fees from users as far away as Australia, and making enough from their activities to live comfortably on the income, had critics wondering if the police crossed the line from investigation to entrapment.

Electronic Frontier Canada, an organization that lobbies for on-line civil rights in Canada, followed the case closely. The organization's president, David Jones, found the circumstances surrounding the bust highly suspect, pointing out at the time that the police were going out of their way to make the crimes seem more serious than they really were. One sysop had a suspicious Australian user who sent a postcard along with his subscription fee, as if to prove that he was indeed from the land down under. It was the kind of thing a Canadian policeman might do to convince a suspicious teenager of his bona fides. Moreover, considering that Australia had, and still has, plenty of its own pirate-software BBSes, it's hard to imagine why this mysterious user would rack up immense long-distance charges to download software from a small Canadian BBS, unless it was a Mountie trying to prove an internation criminal connection. The BBS sysops were mainly students guilty of petty crimes and minor indiscretions. But crossing international boundaries suddenly made the crime more serious.

The Montreal raids would simply have passed into the folklore of the information revolution, except that they came quick on the heels of other raids across Canada. Just six weeks earlier, Mounties raided two adult BBSes in the Vancouver area on suspicion of violating obscenity laws. No charges were laid, but police seized files, computer equipment and detailed information—including photographs—on the BBSes' more than 1700 users. The police were clearly signaling to BBS and Internet users that they could no longer count on the authorities' ignorance of information technology to protect them from the long arm of the law. The cops were getting tough and they wanted the world to know it. However, in their rush to police the apparently lawless on-line frontiers, they disregarded privacy protections that we all take for granted in our homes and places of work.

For traditional legal and judicial authorities, the new technology simply does not enjoy the traditional protections. As lawless as the Internet and BBSes may be, information crime fighting is similarly out of control. The searches of the Montreal BBSes prior to the raids were conducted without the benefit of search warrants, and in normal circumstances,

> Protection against intrusion into one's private affairs protects individuals against unreasonable physical intrusions as well as eavesdropping. One may not physically intrude upon another's private space (i.e., home, hotel room, etc.), tap telephones, read private mail, or engage in certain other surveillance techniques which intrude on an individual's private affairs. To succeed in court, one must demonstrate that an intrusion into a private matter has occurred. If the matter is one which can reasonably be considered public, this claim will fail.[1]

Despite the fact that the surveillance was carried out under false pretenses—the police operated undercover and did not identify themselves as law enforcement officers—on private systems maintained in private homes, with a private list of subscribers, the court ruled that it wasn't an intrusion into a private matter. Thus protections against unwarranted search and seizure did not apply. The law is being applied on the new frontier like a light beer—all the sanctions with none of the protections. It appears that, in the world of on-line law enforcement, the end justifies the means. "In the B.C. case, you have to wonder if the database of personal information on users—including, presumably, logs of who downloaded what—may have been the target in the first place," Jones said. "Because these BBSes were on Fidonet [an international network of BBSes], they also seized an indeterminate number of e-mail messages *in transit*. We have legal protections for letters in Canada Post, but evidently not yet for electronic communications."

II. The long arm of the law

Freedom... privacy... protection from arbitrary search and seizure; these are the issues of the Internet, and they have begun to exert a direct influence on the public policy agenda wherever the information revolution has taken hold. The rancorous debate over who has jurisdiction—and if there *should* be jurisdiction— is a manifestation of the discontinuity between the on-line polity and the off-line world of traditional authority structures and jurisdiction. If the values and ideas expressed on-line were completely consistent with the off-line world, the question would never be asked. Because the flow of information defies authority predicated on geopolitical jurisdictions and notions of property based on concrete commodities, the information revolution has caused a legal dislocation. Laws and sanctions designed to establish ownership over property that you can see and touch, and restrictions on information that once had to *physically* cross borders have become inadequate in the wake of the information revolution. It is no accident, then, that the recent history of government information technology policy—that is, policies to deal with information technologies—has been dominated by an effort to assert control. In effect, wherever the information revolution has established itself, governments have attempted to effect a counter-revolution.

The first salvos of the counter-revolution were fired even before most

North Americans knew what the Internet was. The crackdown on hackers and credit card and telephone fraud launched by American law enforcement officials in the spring of 1990—often generically referred to as "Operation Sun Devil" after the U.S. Secret Service raids of May 8, 1990—had next to nothing to do with the information revolution, and even less to do with the Internet. It was simply a sweep of petty criminals and grifters who used information technology and telecommunications networks to commit their larceny. However, police sometimes cross the line from law enforcement to harassment, and with the distinction obscured by the rapid developments of the information revolution, rights and freedoms were dragged up in Sun Devil's nets.

Above all, Sun Devil was an over-reaction to a mounting fear in the United States that computer criminals were getting out of hand. Fueled by the 1983 Hollywood movie *War Games*, in which a teenaged hacker played by Matthew Broderick inadvertently begins the countdown to Armageddon after breaking into the North American Air Defence Command's computer network, and Clifford Stoll's *The Cuckoo's Egg*, a self-serving account of how he single-handedly (he claims) thwarted the nefarious designs of hackers in the pay of the KGB, anti-hacker panic had reached a crescendo in the American and Canadian press. The information revolution—still embryonic but somehow threatening—and computers in the hands of high-tech delinquents seemed to be agents of corruption. Law enforcement agencies had to act, if only to demonstrate that computer crime was not out of control. A good number of the hackers, phone phreaks and fraud artists arrested at that time probably deserved their fate, but many did not. Honest computer users, writers and business people were caught up in the sting by association with the hacker subculture or because they flaunted the trapping of cyberpunk techno-rebellion; the Sun Devil police investigators didn't seem to care. As would be the case a few years later in Canada, the end justified the means.

By that summer something extraordinary had begun to happen. The questionable legalities of much of the Sun Devil crackdown spawned the beginnings of a political response from the on-line polity itself. The legal offensive of the previous spring had underlined the discontinuity between traditional civil society and the new polity coalescing primarily around BBSes, on-line services and the Internet. To many of the Internet's pioneers, this wasn't simply a question of crime and punishment, but an invasion. It was a frontier war between settlers who only wanted to be left alone, and the forces of power and capital that sought to usurp the information landscape. In the summer of 1990 a varied group of computer industry leaders, civil libertarians and idealists met to found the Electronic Frontier Foundation.

> The EFF founders saw, as the first reporters from the mass media did not, that Sun Devil was not just a hacker bust. The EFF founders agreed that there was a good chance that the future of American democracy could be strongly influenced by the judicial and legislative structures beginning to emerge from cyberspace. The reasons the EFF spoke out regarding [arrested hackers] Acid, Optik and Scorpion as well as Neidorf and Jackson had to do with the assumptions made by the Secret Service about what they could and could not do to citizens.[2]

The EFF was just the first of many advocacy organizations for the on-line polity. Electronic Frontier groups have sprung up in Canada, Australia and Europe—in fact, in almost every technologically-advanced country. Their avowed purpose is to educate the public about civil liberties in cyberspace, clamour for fair treatment in the vague grey area of the law dealing with computer-related crimes and intellectual property in the information age, and fight legislation designed to regulate on-line content. They also serve a far more important and complex role as the core of the on-line polity's resistance to the arbitrary exertion of off-line authority.

Make no mistake, this is a political resistance, pitting the values inherent in the Internet's very structure against the four pillars of traditional authority—law, property, borders and taboo. The interests of the State are incompatible with the free flow of information and a completely decentralized political discourse. What authority can it have when internal affairs become global issues and its proscriptions are unenforceable? The answer is ambiguous at best, but it is clear that the traditional authorities in societies touched by the dramatic economic and social transformations of the information revolution look on the emergence of the on-line polity—a parallel civil society and political discourse—with alarm. Through legislation, police action and proxy spokesmen they have begun a process which, it is hoped, will close the floodgates and roll back the revolution.

III. Property and Theft

Ownership and property are the nexus of the economic and political, where exchange meets notions of control, and it is here that the first battles of the politics of information have been fought. The new economy deals in intangibles, in information, and it's becoming clear that our economy will no longer be driven by a trade in tangible goods, but by a traffic in data. Though the commodification of information is a relatively new phenomenon, the sale of information products is not. Music, stories, movies and news have been bought and sold for millennia. However, they had always been inseparable from their media of exchange. When you bought a song in the past, it was transmitted in the concrete form of a record or an audiocassette, and a story always came in the form of a book.

Information technology has made the medium used to transmit the information irrelevant. Anything that can be digitally encoded—that is, any information product—can be bought and sold as nothing more than a collection of bytes. While this might make the transmission of the product far easier and more immediate than ever before, it has also created a crisis for the business of information.

> The computer is the single most important villain wreaking havoc with the neat lines of intellectual property and control. Turing's original conception of the computer was a universal machine, capable of copying any other. Such a copying capability is fundamental to the computer's operation... An originator of information has no way of controlling the uses to which that information is put or the ways in which it is read. Computers linked to the network can easily copy a message or file and re-transmit it to any number of destinations. The problem is one of scale and speed.[3]

Traditional property protections predicated on the exchange of tangible media are completely inadequate to deal with this flow of disembodied information on the Net. Some, like Net guru Nicholas Negroponte go so far as to argue that copyright is an outmoded "Gutenberg artifact"[4] due for an appointment with the dustbin of history.

Whatever the case, it is clear that the traditional notions of ownership that sustained the exchange of intellectual property—intangible products of the mind—for generations have been thrown into a crisis. The technologies are evolving much more quickly than our laws. The future of our economy is moving toward information production, but without some kind of protection, these products have no value. The problem is that intellectual property exists only as far as the law permits it to exist. It stands to reason that an information economy requires information products, but without adequate legal protections for intellectual property information producers may be discouraged from selling their intangible wares. Unfortunately, the law plods while the information revolution proceeds at breakneck speed.

There is little doubt that the law has to adapt to the information economy. Though legislators and policy makers continue to dither over what the technology means, the really important question is how are we going to adapt to it. Should the law change incrementally, or as a revolution? Without a doubt, it would be easier incrementally, but we may not have time for that. The changes in the information revolution are so profound that we risk creating an unworkable patchwork quilt of law. Moreover, how we protect intellectual property and information products ultimately has important consequences for society as a whole. Copyright law really reflects our society, but what rights and whose rights should be protected? Unfortunately, the crisis is not being resolved legally, but politically.

It isn't enough to simply say that all information products currently protected by intellectual property laws will be covered as before. Intellectual property protection has always been a compromise at best. Symbolic and information products cannot be accumulated or traded in the same way as copper ingots and soft drinks, yet the law protects them in the same way. This compromise has often led to ludicrous situations like the one which occurred when television talk show host David Letterman left NBC television for CBS. The network insisted that all of the material that Letterman and his writers—many of whom joined him at CBS—had developed for the old show was *its* intellectual property. Even Letterman's famous Top Ten lists were deemed to belong to CBS, despite the fact that wags have been producing comical top ten lists for generations. Presumably everyone who compiled such lists around office water coolers was in violation of CBS's copyright. Traditionally, copyright can only be applied to unique content rather than ideas, but in the wake of the information revolution, the distinction between the two can be difficult to see.

Computer criminals like hackers and pirates were the focus of the traditional authorities' first on-line assault because they were such easy targets. Everyone understands theft and trespassing even if they don't quite understand the Net. Because, in the eyes of the law, computer criminals *are* criminals, the governments, lawmakers and police can stride into the electronic frontier like Wyatt Earp and count on widespread public support when they clean up the gaming houses and saloons. However, like the legendary marshal, the traditional authorities are bringing law and order to the information society not so much out of any moral conviction, but to assert and protect their own profits and control. While it may be hard to argue with the results of the legal crackdown on computer crime, the methods and motivation are suspect.

Why, for example should the RCMP spend hundreds, maybe thousands of person-hours, and employ questionable tactics to hunt down a few teenagers for pirating software? Of all computer crimes, software piracy is both the most widespread and mundane. A single set of diskettes can install a program in an unlimited number of computers. However, when you purchase software, you only acquire rights to use one copy. Any duplication, except for back-up purposes, is usually a violation of the licence agreement that you enter into at the time of purchase. Almost anyone who uses a computer probably uses pirated software. The office worker who makes a copy of his company's spreadsheet program so he work on his home computer on weekends probably isn't even aware that he's breaking the law. Many computer novices simply accept offers of software copies from their friends and co-workers without considering the legal ramifications. Unlike hacking, which usually requires formidable technical skills, and whose perpetrators—except for a few high-profile cases—are rarely caught, piracy is an easy crime. It's so easy in fact that it is the information economy's equivalent of jaywalking; everyone does it, and no one except a few powerful corporations are really hurt. And even that is questionable.

The Canadian Alliance Against Software Theft, a Canadian computer industry organization representing software giants like Microsoft and Novell, estimates that the industry lost revenues of $17.3 billion worldwide, and $316 million in Canada alone to piracy. However, this estimate is probably overblown. No one has yet produced numbers to show how many people who use pirated software would actually have purchased it if they had to pay for it. Surely, a computer game that was copied, played once and then forgotten doesn't represent a lost sale. Indeed, because few of us openly advertise our piratical activities, CAAST's $17.3 billion figure can only be an estimate, and is based on the full retail price of software products without considering educational and corporate discounts. Like leakage from a tap, or the inevitable photocopies of popular magazine articles, some piracy is probably an inevitable part of doing business in the information economy.

For the software industry, piracy is the most heinous crime in the information economy. Industry rhetoric is typically dire, warning that copyright violations will bring the information economy to its knees, while neglecting to mention the vast profits software companies continue to enjoy. "Piracy works against everyone's interests," said Stanley Weiss, general manager of Novell Canada and a member of CAAST's board of directors. "Our material costs are minuscule, but this is a tremendously R&D oriented business. If we don't have the profits to cover development, then we simply won't be able to produce better productivity enhancing tools."

Not the inevitability of piracy, or massively over-priced products, or the fact that the software industry considerably overstates its expenses alters the fact that it is theft. However, in the brave new world of information, the concept of intellectual property has acquired a distinctly medieval feel. In the absence of legal precedent and clear statutes, might makes copyright; the more powerful you are, the easier it is to enforce your claim to ownership or exclusivity. While the Mounties in Canada and the secret service in the United States pursue software pirates—though, in all fairness, no one has shown much inclination to bring the average computer user with a pirated copy of *Doom* to justice—corporate information producers have been able to take what they want with near-impunity.

In the winter and spring of 1996 several newspaper organizations began a campaign to wrest control of electronic rights from freelance journalists and writers. Across the United States and Canada publishers informed their contributors that, henceforth, they would claim the right to reproduce any of the material they had paid for in electronic media like the Internet and commercial database services. Until then, it was understood throughout the publishing industry that, in the absence of a contract stating otherwise, a freelancer only sold the right to reproduce his material once in one publication. Publishers had been violating this right for years by including freelancers' work in commercial databases like Infomart-Dialog without their consent.[5] As the newspaper industry began to look to the Net, publishers realized that they would have to legalize their long-standing practices.

In March 1996, Southam News, the largest newspaper chain in Canada, informed freelancers that, if they wanted to keep writing for Southam's newspapers, they would have to sign a new contract granting the company the right to use their material in any newspaper in the chain and reproduce it "by any means or technology, as part of the database of the relevant newspaper or newspapers or in products derived from it."[6] In effect, the company was asking its writers to sign a blank cheque, without any assurance of additional compensation, or face the prospect of never working for the company again. Southam's rationale was that the electronic publications that the company was developing were simply "extensions" of the newspaper itself, and thus did not warrant additional compensation. Not surprisingly, most of the writers gave in after putting up a brief, largely ineffectual fight. The contract presented to freelancers the following year went some way toward clarifying the situation with a detailed schedule of reprint fees, but the company still refuses to grant additional compensation for electronic rights.

Ironically, Southam itself became the victim of information piracy a few months later, when the *Toronto Star* (not a Southam paper) began running Canadian Press wire stories that had originated at Southam on its Web site. Using the same logic that Southam had used on its freelancers, *The Star* argued that its Web site was simply an extension of the print version of the newspaper, and was thus included in its licence for CP material. Southam disagreed, threatened to pull out of CP—leaving the wire service a shell of its former self—and ultimately prevailed. As long as electronic intellectual property rights remain a grey area, copyright will belong to whoever has the might to extort or protect it.

It's not a matter of information technology changing our lives sometime in the distant future. The revolution is happening now, and the need to define such basic issues as protections for intellectual property and copyright in an environment like the Internet has become an urgent matter. A new generation of technologically savvy lawyers might be part of the answer, but as participants in a global information economy it is imperative that we start to deal with the legal implications of the information revolution right now. The issue is political, and whether we meet the challenge will depend on the political will to guarantee *equal* information rights. "It's difficult for people to adjust their mind-set if they're used to a goods-based economy, but we're moving from a goods, to a services, to an information economy," said Sunny Handa, one of Canada's leading intellectual property experts. "I don't think that government has come to grips with the fact that this technology is what our country is going to be about in a few years."

The real answer will be an understanding of intellectual property and copyright that takes the realities of the information revolution into account. It isn't enough to rely on outmoded ideas that benefit the traditional accumulators and controllers of information. The promise of empowerment and enfranchisement held out by the information revolution, and the Net it particular, is predicated on access to information. However, finding an equitable path between these two trends is no easy matter. Until individual information providers, including journalists, artists, Web designers and shareware authors, can count on the kind of legal and judicial support freely extended to corporate software developers and newspaper chains, access to and control of information will continue to be unequal. And that is exactly the crux of the politics of information. The battle for control of the Net is a battle for control, to stay the tide against information freedom. Traditional authority rests on the shoulders of gatekeepers who guard morality, taboo and the limits of speech. For these authorities, power is expressed as limits, and the extension and maintenance of power requires the extension of those limits to the vistas of information.

IV. The Gatekeeper

It's tempting to see Sol Littman as one of the good guys. A long-time activist for social justice and tolerance, Littman has been one of Canada's most prominent leaders in the fight against anti-Semitism for three decades. He has worked with the Anti-Defamation League in the United States, published books exposing the admission of Nazi war criminals to Canada after the Second World War, and campaigned tirelessly to bring unrepentant Nazis to justice. And as Canadian director of the Simon Wiesenthal Center, he is one of the Canadian Jewish community's most important spokesmen. Like many of his generation,

Littman was completely taken by surprise by the information revolution and the growth of the Internet. Due to his inability or unwillingness to understand the profound social impact of the new information technologies, not to mention how they work, Littman has become one of the most vocal and influential advocates of strict Internet regulation and opponents of information freedom in North America.

In May 1995 I was invited to speak on a panel dealing with hate propaganda on the Internet at the Canadian Jewish Congress's annual plenary conference in Montreal. Joining me on the panel were Ken McVay, who was then just getting the Nizkor Project off the ground, and the CJC's Bernie Farber. The audience was a broad cross section of the Canadian Jewish community—professionals, senior citizens, students, young professionals and Holocaust survivors. They each had widely different experiences of the Net, but all shared a common experience of bigotry and anti-Semitism... and they all wanted to know what can be done about racists and neo-Nazis on the Net. McVay and I, who had just met for the first time minutes before, were in complete agreement. Without wishing to trivialize the fears of people who had had first-hand experience with the horrors of the Holocaust we calmly explained that on-line hate propagandists were neither numerous nor much of a threat to the security of the Jewish community. They warranted concern, but definitely not panic.

We made our points calmly and rationally, made what I thought was a compelling case for on-line anti-racist activism without the need for government regulation. McVay and I probably got through to most of the audience, but when questions were invited from the floor, it became abundantly clear that the Net wasn't going to get off so easily. Littman, from the back row, stood up not so much to ask a question, but to harangue the audience on the danger posed by hate propagandists. He trotted out assertions that have become recurring themes in his crusade to bring government regulation to the Internet—you can't reason with racists; trying to reason with racists only legitimizes their ideas; the Net is a cesspool of sin and corruption.

It occurred to me then that Littman was grandstanding, using the panel to advance his own theories about freedom of speech and access to information, and within a month, I was proved right. In June, the Wiesenthal Center released a discussion document (that's how Littman describes it) entitled *The Need for*

Regulation on the Information Highway, to Canadian legislators, the press and the Canadian Radio-Television and Telecommunications Commission—the government body responsible for regulating the broadcasting and telecommunications industries—calling for a legal crackdown on the Net to combat hate propaganda. Among the document's recommendations were broad additions to Canada's already strict anti-hate laws and an "international conference to arrive at uniform, world-wide regulation."[7]

Littman later admitted that, at the time he wrote *The Need for Regulation...* he had never actually *seen* the Internet. The whole report was based on the work of a single research assistant. In fact, it is questionable whether he had ever even used a computer, since the document was prepared on an obsolescent IBM Selectric typewriter. (Significantly, *The Need for Regulation on the Information Highway* is not available in an electronic form anywhere.)

The Wiesenthal Center report left no doubt how far it believed the Internet should be regulated. It was the most radically restrictive plan for government control of Internet content ever proposed in a western democracy. The Center proposed that all Internet communication with the exception of person-to-person electronic mail be defined as broadcasting, thus placing it under the CRTC broadcast regulations.[8] Needless to say, if such a plan had ever been implemented, it would have thrown an impenetrable pall over the use of the Internet in Canada. When asked if this covered text messages on FTP sites and posts to Usenet newsgroups, Littman was quick to point out that he intended to censor "any public messages, nothing should be out of bounds." Littman had no trouble with the implications of blanket censorship. In his mind, the Internet is not composed of people, but of millions of small-scale broadcasters. He was overwhelmed by the mundane technology of computers, and because Internet communication is mediated electronically, he believed that it is not casual conversation in a new medium, but broadcasting. Ironically, while he contended that individual users are all broadcasters—he believed that Internet service providers are as well. In Littman's mind, *everyone* who communicates via computer networks is a broadcaster, and should thus be required to apply for a licence. Without considering how Internet service providers would absorb the cost of undergoing the CRTC's lengthy licence hearing process, the Wiesenthal Center placed the responsibility for monitoring and regulating several hundred providers in the commission's lap. In any event, with telecommunications monopolies, deregulation of the telephone industry, and Canadian content requirements to worry about, it's unlikely that the CRTC would have welcomed the responsibility.

Littman's proposals had little to do with reality to start with. The purpose of the document was to create a climate of fear, to inspire panic in otherwise reasonable people whenever they looked at a computer. Littman chose his words and illustrations carefully, prefacing the report with a quote from an inflammatory message from a known white supremacist, while neglecting to mention that it did not violate Canadian hate laws and that, in any case, was posted by an American user, far beyond the reach of Canadian courts. The document actually *under*-estimated the number of on-line ultra-right groups at fifty,[9] but while the number sounded threatening the Simon Wiesenthal Center forgot to mention was that most of these groups consist of one or two people, and that, among the tens of millions of Internet users, they are a microscopic minority at best. "We're not a panicky organization," Littman said. "We're responding to something we see as a real problem... We just don't want to see the Internet collapse under its own follies." Rather, he would be happy to see the Net collapse under the weight of off-line panic.

The document was released with great fanfare and ceremony on Parliament Hill in Ottawa, Littman's stunt reaped great media coverage dividends, and two years later is still often cited as the definitive statement on the need to regulate hate propaganda on the Internet. Canadian Jewish Congress president Bernie Farber raised the red flag of panic, referring to the report on radio phone-in shows in the summer of 1995. If a plan to regulate the Net for unpleasant content continues to exist in Canada, it is largely because of the abject fear that Littman put in the hearts of Jewish community leaders. Though his recommendations have not been acted on, *The Need for Regulation on the Information Highway* did successfully accomplish two things—it created an atmosphere of panic and, using that fear, consolidated the Wiesenthal Center's position as the Canadian Jewish community's leading anti-Nazi organization. Littman's power and prestige is predicated on the existence of an immediate neo-Nazi danger, and though it has become clear that the ultra-right's on-line presence, as objectionable as it may be, does not constitute much of a social threat, he has gone out of his way to create the appearance of one.

At a Jewish community event where I spoke about on-line hate propagandists in the fall of 1995, an elderly Holocaust survivor pulled me aside and said "they're coming again, and we have to do something to stop them." She was in a panic about a resurgent Nazi movement that Littman had promised was once again on the march, but this time in cyberspace. Whether these Nazis were real or virtual didn't matter much to this survivor, the mere *idea* of them was enough to scare her out of her wits. "We should shut this Internet down," she said. "Sol Littman is right." It occurred to me that the elderly woman probably believed that unsuspecting users were being jumped and attacked by Nazi thugs every time they logged on to the Internet. After my presentation, in which I demonstrated the various Internet technologies and gave the audience a tour of neo-Nazi Web sites and Usenet, the survivor approached me and said "I didn't think it was like that... it isn't as bad as I thought it was." More to the point, it wasn't as bad as she had been led to believe it was.

The Holocaust and the threat of Nazi violence are Littman's franchise. His whole message is paternalism at its worst—there are nasty ideas out there that society has to be protected from, and *he* is the one who will do the protecting. What seems to bother him the most is the way in which the Net either subverts or circumvents traditional authority structures. His essay "Some Thoughts on the Regulation of Cyberspace," appended to the 1995 report, is particularly revealing:

> The cyberspace information-giver need not list his qualifications, display his degrees, prove his competency, supply proof of his membership in a professional society or provide his bibliography... The necessary props of scholarship are obliterated, expertise goes out the window.[10]

In short, Littman's principal complaint is not with *content*, but with the potential of the Internet to enfranchise groups and individuals that were silenced by the rigid patterns of authority and prestige prior to the information revolution. His problem is with the apparent lack of authority, a manifestation of the Net's decentralized discourse and distributed structure. Dismissing Internet civil libertarians as children—implying, of course, that they require steadying adult supervision from people like Sol Littman—he attacks the very interactivity that provides activists like Ken McVay with the opportunity to strike back at hate. Despite the Nizkor Projects successes, Littman believes that it doesn't go far enough, and that McVay's tactics are an invitation to disorder. In his mind, meeting the Neo-Nazis head-on is "a dangerous invitation to digital vigilantism and promiscuous computer violence"[11] that will do little but turn the Net into an electronic frontier town where the lynch mob is justice.

What people like Littman fear is that on the Net authority will pass from traditional power centres to the users themselves. He is a traditional gatekeeper, seeking to preserve the prerogatives of traditional authority in a medium that he simply does not understand. Expressing parental concern that we "need to protect children against smut and limit the spread of hate literature," he appeals to instinct to buttress an argument that cannot be supported by reason, and which flies in the face of the information revolution.

V. Protect the Children

Shortly before Christmas in 1993, a fifteen-year-old boy in Laval, Quebec lost part of his hand when a pipe bomb he had been making went off. Teenage boys have been building explosive devices with pipes, charcoal, and sulfur and saltpeter from high school science labs and pharmacies for decades. As tragic as it was, this incident was not particularly remarkable except for one thing—the boy claimed that the bomb making instructions had come from a document called "Anarchy for Fun and Profit" downloaded from a local BBS. Within days of the revelation, the story, which would have rated the inside pages of a daily newspaper, appeared under the headline "Computer data can be dangerous" on the front page of the Montreal *Gazette*, one of Canada's major daily newspapers.

Tina Crossfield, a suburban mother and a graduate student at Montreal's Concordia University, emerged as the standard bearer for one of the first cyberspace regulation campaigns in Canada after she found that her son had also downloaded the 'anarchy' file. She quickly became a media darling, appearing on radio programs and being quoted in print media across Canada. Her cause—to make cyberspace safe for her children—struck a chord among parents who, like her, looked on computers, BBSes and the Internet as nothing less than mortal danger in technological dress. "Something has to be done," she said at the time, arguing, like Littman would a year later, that all networking systems from BBSes to the Internet should be tightly controlled by new laws. "The computer phenomenon is such a new and complex thing. TV has been around for a long time and the regulations on violence and sex have built up over a number of years. It's time to apply some of that to the BBSes and networks."

Crossfield's crusade had little effect except to bring the question of regulating computer-mediated communications to the front pages. It was one of the first gusts of a chilling wind that began to blow through the on-line community. With "protect the children" as their slogan, the forces of regulation and control began a crusade against dangerous content and smut that has profoundly affected the public perception of the Internet as it emerged as the dominant medium of the information revolution.

From that day onward the question of *when* the boom would drop became the on-line polity's principal preoccupation. Internet users and BBS system operators argued over what to do, but there was little doubt that it was going to happen—and sooner rather than later. "That's the biggest rumour on the Net," said Montreal sysop Vince Vallée at the time. "It's been around for years, and it gets repeated whenever people start to feel insecure. If there *is* going to be a crackdown, the guidelines have to come from the people who know the most about the medium, not some senator in the U.S. or Canadian M.P. Regulation may be inevitable, but people have to know the big picture. What we need is some hard legislation to define our rights, so the on-line community itself can define its responsibilities."

In fact, the big push for regulation did come from an American senator. Clutching a binder packed with pornographic images—the bluebook of the information revolution—in one hand, Nebraska senator James Exon strode the floor of the U.S. Senate, calling for a war on Internet indecency. His Communications Decency Act, first introduced to the Senate in 1994, would prohibit "any comment, request, suggestion, proposal, image, or other communication which is obscene, lewd, lascivious, filthy, or indecent..." on the Internet. Perhaps American legislators were not ready at that time to tackle a technological issue that most of their constituents did not yet fully understand, and Exon's proposed law, co-sponsored by Senator Daniel Coats languished on the legislative agenda until the spring of 1995. By then the Internet had become big news. Internet-related companies were raking in bundles of cash in public stock offerings, and Web addresses were appearing on advertisements for sneakers and soda-pop.

Once the domain of a select group of engineers and graduate students, the

Internet had become a mass medium, and with that it had come under the close scrutiny of gatekeepers like Exon and Coats and the traditional arbiters of morality. The now-infamous *Time* magazine "Cyberporn" cover story of July 3, 1995 was both an indication of and a catalyst for the panic that somehow forced the CDA through both houses of the U.S. Congress. Accompanied by a cover illustration showing a wide-eyed young boy staring into the lewd phosphor glow of a computer screen, Philip Elmer-Dewitt's article focused on a just-released study that purported to show how bad the problem of on-line pornography really was. According to the article, the findings of an "exhaustive" study on on-line pornography by a research team at Carnegie Mellon University were "sure to pour fuel on an already explosive debate."[12] Elmer-Dewitt was partly right. "Marketing Pornography on the Information Superhighway" brought the debate over Internet regulation from a slow simmer to a fast boil; both Exon and Coats used it to revive the CDA, which had begun to lose momentum during the slow summer congressional session.

The study was not exhaustive, or the work of a research team, or sanctioned or funded by Carnegie Mellon University. It didn't even have much to do with the information superhighway. "Marketing Pornography" was a seriously flawed research paper by an undergraduate named Marty Rimm that focused almost exclusively on the tastes of users at a handful of adult BBSes. Rimm did take a sidelong glance at the Internet, but treated it principally as a marketing adjunct for the BBSes. It didn't take long for Internet civil libertarians, serious academics and journalists to pick it apart. Where Rimm disclosed his methodology—an essential part of any serious scholarly research that the author usually tended to ignore—it was found to be nonsensical. He claimed to have used sophisticated dictionary software to categorize pornographic images that turned out to be simple database scripts inadequate to the purported task. Rimm conflated findings from adult BBSes with the Internet, which is akin to basing a study of the general public's drinking habits on liquor store inventories.

Most importantly, Rimm's statistical findings were either insupportable or just plain wrong. For example, Rimm maintained that, in the alt.binaries Usenet newsgroups—forums where users post digital images—83.5 percent of the total posts were pornographic. What Rimm failed to mention was that individual images are often broken into as many as fifty segments, depending on the image's size, each posted in sequence to the group. If pornographic pictures tended to be larger than non-pornographic images, fewer actual *images* could easily account for more *posts*, moreover, Rimm had based his conclusions on written descriptions rather than the images themselves and had only counted posts in a handful of newsgroups.[13] Despite Elmer-Dewitt's overheated prose, Rimm only proved the obvious—that some computer users made available and consumed pornographic images on the Internet. His report was soon discredited on the Net, in the press and at universities, but the harm had been done. As spurious as Rimm's findings were, they simply confirmed the worst fears of a public trying to make sense of the opportunities and challenges of the information revolution.

Elmer-Dewitt's article ran two weeks after the Senate approved the CDA as part of the omnibus Telecommunications Act by an 84-16 margin. While Exon had received an advance copy of Rimm's report, and had used it to advance its case, the *Time* cover story played no part in the debate. It could not have come at a better time for the pro-CDA forces in the House of Representatives. Though most Americans had previously been indifferent to on-line pornography, the controversy surrounding the article brought Cyberporn to their attention and precipitated a media feeding frenzy. In the summer and fall of 1995, you could not turn on the television without seeing dirty pictures—tastefully censored for prime time—scroll by in a browser window while pundits discussed the prurient dangers of the Internet. In the rigid apple-pie-and-family-values atmosphere following 1994's 'conservative revolution,' cyberspace sin became a hot topic—and congressmen facing re-election in less than a year could not exactly put themselves in a position where they might appear to be soft on sin, or pro-pornography, despite whatever misgivings they might have had about the Communications Decency Act.

The outcome was never in doubt. The CDA was approved in the House and signed into law by president Bill Clinton in February, 1996. Then something interesting happened. A broad coalition of off-line and on-line civil libertarians began an aggressive campaign to stop the law from being implemented. Web sites all over the Net went black in protest for 48 hours beginning on February 8, and the American Civil Liberties Union, together with a number of other groups, immediately launched a constitutional court challenge. The ACLU charged that the CDA applied a far harsher standard of decency to the Net that was applied in any other medium. Not only did the act ban illegal obscene material on the Internet, but in the interests of protecting the children, it outlawed expression and speech which enjoy constitutional protection in almost *every other medium*. Information on sexually-transmitted diseases, contraception and abortion, legitimate artistic and literary depictions and references to sexuality, even passionate messages exchanged by teenaged lovers—all of these were forbidden under the Communications Decency Act.

On June 11, 1996, almost exactly a year after it had been passed in the U.S. Senate, the ACLU's campaign to have it stricken from the books was successful. The three judges of the District Court for the Eastern District of Pennsylvania did something that neither Exon nor Coats had bothered to do—they explored the Net. The decision (ACLU v. Reno) reflected that fact, and in his written decision, Judge Dalzell observed...

> As the most participatory form of mass speech yet developed, the Internet deserves the highest protection from governmental intrusion.
>
> True it is that many find some of the speech on the Internet to be offensive, and amid the din of cyberspace many hear discordant voices that they regard as indecent. The absence of governmental regulation of Internet content has unquestionably produced a kind of chaos, but as one of plaintiffs' experts put it with such resonance at the hearing: *What achieved success was the very chaos that the Internet is. The strength of the Internet is that chaos.*
>
> Just as the strength of the Internet is chaos, so the strength of our liberty depends upon the chaos and cacophony of the unfettered speech the First Amendment protects.[14]

Not only is the decision the most important statement yet of the need to *protect* rather than *restrict* the chaotic conversation of the Internet, but the panel of judges went farther than anyone else in discrediting the gatekeepers' contention that special regulations are required to 'protect the children.'

> This rationale, however, is as dangerous as it is compelling. Laws regulating speech for the protection of children have no limiting principle, and a well-intentioned law restricting protected speech on the basis of its content is, nevertheless, state-sponsored censorship.

> Regulations that "drive certain ideas or viewpoints from the marketplace" for children's benefit, risk destroying the very "political system and cultural life," that they will inherit when they come of age.[15]

Dalzell observed that the CDA "will almost certainly fail to accomplish the Government's interest in shielding children from pornography on the Internet" because at least half of the Internet's traffic originates beyond the jurisdiction of American laws. Of course, with the U.S. government's historic propensity to extra-territorial enforcement of its domestic laws (witness the Helms-Burton Act), this is not entirely true.

Most damaging to the regulationists' cause, however, was the court's contention that the CDA was simply *unnecessary* to protect children. The belief that the Internet was *so* bad that it had to be tightly controlled lest the vilest forms of pornography twist the impressionable minds of our children is a central tenet of the pro-regulation discourse.

> But pornography is different on the computer networks. You can obtain it in the privacy of your home—without having to walk into a seedy bookstore or movie house. You can download only those things that turn you on, rather than buy an entire magazine or video... The great fear of parents and teachers, of course, is not that college students will find this stuff but that it will fall into the hands of those much younger—including some, perhaps, who are not emotionally prepared to make sense of what they see.[16]

However, the court didn't quite see it that way. Having taken the time to explore the Net, the panel of judges realized that the Internet is an environment not unlike a village square or a suburban yard, and that parents who supervised their children in such real environments should be equally prepared to watch their play in virtual space. Moreover, with the development of parental control software that can limit youngsters' electronic travels technologically, parents can protect their children without stepping on the rights of *adult* users.

> Parents, too, have options available to them. As we learned at the hearing, parents can install blocking software on their home computers,

> or they can subscribe to commercial online services that provide parental controls. It is quite clear that powerful market forces are at work to expand parental options to deal with these legitimate concerns. More fundamentally, parents can supervise their children's use of the Internet or deny their children the opportunity to participate in the medium until they reach an appropriate age.[17]

Rather than reduce the Net to some homogenized Disney World of family values, the court said simply that it is, after all, the parents' responsibility to supervise their children's wanderings on-line as well as off-line.

Of course there is pornography on the Internet, and some of it is pretty raunchy, but it is significant that the CDA attempted to apply a vague, indefinite standard of decency to the Net rather than enforce the generally accepted legal standards of obscenity. Obscenity is not protected under the constitutional right to free speech, but its standard is both strict and narrow, and generally excludes written material. Gareth Sansom, a researcher for the Canadian Government's Information Highway Advisory Council, concluded in a report that the vast majority of pornographic material on the Net doesn't even violate Canada's far tougher obscenity laws.[18] With pornography, as with everything else, there is nothing on-line that doesn't exist off-line, and almost everything that you can download has already appeared in print or on legally-available videocassettes. If the CDA's standards were applied to *all* media, it's unlikely that even prime-time television programs dealing with adult themes and literary magazines would have passed the decency grade.

The CDA was not about decency any more than it was about protecting children. It was about power. As with Littman's anti-Nazi campaign in Canada, it was about fear. Though they have heard about it on television, read about it in magazines and talked about it around the office water cooler, most North Americans know little about the Net. Like Exon, they find it strange, foreign and vaguely threatening. The infinite possibilities of the new medium can equally be new threats, and supported by Rimm's sensationalist pseudo-science and Elmer-Dewitt's hyperbolic bombast, many see it as something to be feared. That works to the advantage of legislators who want, above all, to extend their power to a new frontier.

The subtext of the regulation debate isn't social or cultural, but political.

The big question isn't how to protect children or eliminate racism, but who will control the Net. The regulationists have woken up and discovered to their horror that one of the most important media for communication and the exchange of ideas is not run by a government agency, but by the users themselves. To them, the idea of leaving governance of the Net to the on-line community and 'frontier justice' is unconscionable, no matter how well it works. Information is power, and the drive to control information technologies is nothing more than a power grab.

VI. Tightening the Noose

While the CDA's direct assault on the Net failed, momentarily at least, in Canada the traditional authorities have pursued the same goals by stealth. The Canadian political culture is predicated on the doctrine of elite accommodation, and rather than grabbing control of the Net in a frontal assault of legislation they have pursued a strategy of co-opting leading interests in government policy. While people like Littman and law enforcement officials have raised the alarm and kept the issue of regulating Internet content on the public policy agenda, the Canadian government has quietly set its own agenda through consultations and meetings of the Information Highway Advisory Council, a government body created under the aegis of Industry Canada to establish the country's Internet policy. IHAC studied a wide range of issues, from universal access, to satellite TV's implications for Canadian media companies, and it has made a valuable contribution to the public debate on information highway policy. Nevertheless, it was clear from the beginning that an important part of its mandate was law and order on the electronic frontier, and Industry Canada—the ministry IHAC answered to—has taken that to heart.

In September 1996, Industry Canada held a series of meetings in Montreal, Toronto and Vancouver to sound out Canadian Internet experts and entrepreneurs on the pressing question of legal liability on the electronic frontier. You'd think that when the federal government held *public* consultations on an important issue, they might make a point of inviting people to attend, but the meetings were in reality open to a select group. The only people who were informed about the meetings were the handful of Internet service providers who make up the Canadian Association of Internet Providers, Industry Canada's legal consultants, and the government itself. The list of people who *weren't* invited read like a who's who of Canadian Internet experts: there was Sunny Handa, co-author of *Getting Canada Online* and one the country's leading experts on intellectual property law and the Internet, and David Masse, president of l'Association Québecoise pour le développement de l'informatique juridique (The Quebec Association for the Development of Juridical Information Technology). Most surprising of all, Electronic Frontier Canada, the most important organization in Canada for the preservation of on-line rights, only heard about the meetings second-hand.

Despite the fact that the EFC's Jones was never informed of the meetings—he heard about them from Masse, who heard about them from someone else—government officials insisted that his organization was specifically invited. Despite Industry Canada's protests that the meetings weren't *supposed* to be hush-hush, it was hard to shake the feeling that the ministry didn't want to invite anyone who might ask tough questions. There was no mention of the consultations on Industry Canada's homepage, the one place you might expect Internet experts would be likely to look. The government stacked the deck to get the *right* people for its accommodation. "They've gone about this in the most clumsy way possible," Masse said. "When I raised the issue, Industry Canada told me 'we didn't have to consult anyone, anyway.'"

The meetings were held following recommendations in IHAC's 1995 report that the government should consult with legal and Internet experts to clarify the question of legal liability on the Net. The consultations focused on questions of copyright, control of offensive content and privacy. The problem is that if someone violates copyright or breaks the obscenity or hate laws, it's not quite clear who is liable. For example, do we prosecute an Internet service provider for allowing one of its users to maintain an archive of pirated software on its system? Who is liable if child pornography from Finland is available on one of the 10,000 Usenet newsgroups accessible in Canada?

There was reportedly some discussion that this responsibility might fall to the individual Internet service providers. Some of the larger Canadian ISPs had already made efforts to filter out potentially obscene newsgroups, but even setting aside the question of whether this is necessary—only a tiny fraction of the pornographic material on the Net actually violates Canadian obscenity laws—you have to wonder if it's even practical to do this on any large scale. The job would be so big, in fact, that small local operations simply wouldn't have the resources to do it. "Internet access is a very competitive business," Masse said. "The big players would love to eat up market share, and it's possible that these players could be set to dominate the market. We could end up with requirements that are so expensive that they will penalize the small operators."

That would suit the Canadian government just fine. The ISP industry is far too disorganized for government tastes. CAIP itself was set up largely at the pleasure of Industry Canada, doubtless as a nod to the tradition of elite accommodation. The ISPs are the railroads of this generation, and just as the Canadian government of the 19th century quashed competition and helped consolidate the rail industry into two large corporations—one public and one private—it now seems willing to deal with the Internet as one monolithic corporate interest. It's not surprising, then, that almost all of the smaller Internet providers were left out of the loop, despite the fact that the consultations dealt with issues of importance to *everyone* in the business. Indeed, a representative of only one ISP attended the meeting in Montreal. He had been informed of the consultations by Masse himself, but many others—the majority, by Masse's count—were completely unaware that the consultations were taking place. Significantly, representatives of major telecommunications companies like Bell Canada and cable suppliers were in attendance. The exclusion of so many important voices has profound implications for the future development of the network infrastructure which Canadian businesses will need to participate in the global information economy. It may be important to set the legal ground rules for the Internet—that has yet to be established—but the government's consultation process, such as it was, seemed to be more an attempt to consolidate the industry and serve notice to smaller providers that they will soon be forced to comply with content standard. The "Content-Related Internet Liability Study" is due to be completed in 1997, but the effects of the Canadian government's heavy-handed tactics are already being felt.

On November 1, 1996, CAIP announced a voluntary code of conduct "in response to increasing concerns over content on the Internet." Its members agreed to cooperate with law-enforcement authorities, that they will not knowingly host illegal content on their servers, and that they would take steps to investigate any complaints about such content or network abuse. On the following Monday, November 4, police in Kirkland Lake, Ontario, announced the arrest of a man accused of possession of child pornography and participation in an international pedophile ring on the Internet. Talk about good timing.

In fact, there was little evidence that the 'ring' was anything more than a casual exchange of electronic mail. The accused did appear to possess child pornography—and was thus in violation of Canadian law—but the Internet connection was a bit of a stretch. It was a stretch that worked in the authorities' interests. Nothing better guarantees high profile than an Internet connection, and nothing assures increased police funding better than a high-profile case. All of a sudden, child pornography on the Internet was a hot issue in Canada, and CAIP couldn't have better planned the release of its code of conduct if it had tried. The Association tried to distance itself from the Kirkland Lake case, but privately, some members admitted that they had expected "the shit to hit the fan" in November, and felt that they should get something on paper that could "protect our butts." It's clear that police had been cranking up the pressure on the ISPs for some time. The child-pornography arrest—which was actually made some weeks before the November 4 announcement—just made it public. "It was amazingly good timing for us," said Association president Ken Fockler. "We've been talking about a code of conduct for a couple of years—before CAIP was even formed. We had discussed something like a 'Good Housekeeping' seal of approval. I attended the police chiefs' association's meeting in Ottawa in August, and talked to people in the RCMP and the Ontario Provincial Police about what we could do. They were concerned about child pornography on the Internet, so it showed we were on the right track."

More significantly, the rising pressure from police and government corralled the Internet industry and forced the ISPs into a pen. The Association's members were rattled by the hearings, the arrests, and none-too-subtle hints from the law enforcement community. Despite the successful court challenge to the CDA in the United States the previous summer there was a consensus that the Industry should act before the boom fell in Canada. "A lot of ISPs feel concern about content on the Internet, privacy and issues like that," said Ron Kawchuk, CAIP's executive director. "There was concern that the government might try to legislate regulations. I think the industry decided that, if there are going to be standards, then it should set its own."

The code received mixed reviews. Not surprisingly, Littman praised CAIP, but civil libertarians were less enthusiastic, because while the code was essentially common sense, it left a number of issues dangerously vague. It was more of a public relations exercise than an attempt to establish clear and reasonable industry standards. CAIP's members want to convince the public that they're good corporate citizens with good family values. While pointing out that it had no objection to a code of conduct *in principle*, EFC was concerned that, by leaving it vague, CAIP was doing little to protect the rights of individual users. The Association did provide an accompanying commentary on its Web site, but did little to clear up the confusion. On one hand the code clearly stated that "privacy is of fundamental importance," but in the commentary, it was made conditional, "taking into account the relative sensitivity of each type of information." CAIP obviously doesn't want to say that anything is 100 percent private, in case they have to cooperate with the police, but to measure the relative sensitivity of information, they'll have to look at it.

The code was supposed to be a living document, a work in progress that will evolve as the Association receives input from the Internet community. "It's not the easiest thing in the world to figure out what to do and how to do it," Kawchuk said. "This is really just the beginning of a process. In fact, you have to wonder if you can have a code for just one country, or if we need international cooperation with industry associations in other parts of the world." The problem is that the process points to greater regulation and government intrusion into on-line content. The goal is to bring the Net in line by co-opting it at the point of access. International cooperation promises to bring more, not less regulation, with a compromise settling for the lowest common denominator of acceptable content and conduct. And when a global agreement is reached—inevitably through the exertions of the state—it's unlikely that it will be merely voluntary.

VII. Subversive Bits

On March 14, 1995 then-CRTC Chairman Keith Spicer announced a new policy that will require all television programs to be rated for violent or sexual content, and all televisions sold in Canada to be equipped with a V-chip that will let viewers block objectionable content. The V-chip has been discussed in the United States for several years, but the Canadian government was the first to make it public policy. In a way, the announcement was a fitting dénouement to Spicer's career as Canada's top telecommunications policy maker. The V-chip, for all its technological promise, is also a fitting metaphor for the rudderless course taken by the commission—and the government as a whole—during the first part of the information revolution. Here is a whiz-bang electronic gadget that few people really want, and that no one has proved we really need. It solves a problem that may not be a problem, or which can be solved with far simpler, though low-tech methods that don't add an extra cost to new televisions. The V-chip is a triumph of form over substance. If it didn't have a name straight out of Tom Swift, or emerge at a time when the public is particularly sensitive to smut and violence, it would probably languish in the obscurity it deserves.

However, Spicer had to do something—anything—to show that the toothless, increasingly irrelevant CRTC is still capable of leading public policy on telecommunications, and the government had to do what it could to show that, in terms of information technology policy, it is still in charge. After six and a half years in the commission's driver seat—a time when the information revolution took to the high road of the information highway, when some of the most important economic and industrial issues facing our civilization involve the much anticipated convergence of telecommunications and computers—Spicer's legacy is a television rating system that will allow parents to keep their teenagers from watching *Baywatch*.

As our society has become increasingly wired, the commission's role and responsibilities have become ever more important. During Spicer's mandate the CRTC has faded into near-irrelevance. In November 1995 Bell Canada, the supposedly regulated monopoly that controls the local phone systems in Quebec and Ontario, unilaterally hiked the rates it charged to Internet service providers for Centrex III telephone lines, but the commission just sat back and watched. Bell ultimately backed down, conceding that the tariff on which the rate increase was based didn't really apply to ISPs.

The most disturbing part of the incident wasn't that the telecommunications giant was trying to make more money off the backs of independent Internet service providers—Bell has every right to make a legal buck wherever it can—it was the CRTC's position. The commission maintained all along that Bell had acted perfectly within the regulations that governed telecommunications. It took an intervention by the ISPs themselves to prove otherwise. In effect, a group of independent entrepreneurs, whose livelihood was threatened by Bell's illegal rate increase, had to read the regulations back to the CRTC. The commission, which is supposed to know what its regulations mean and enforce basic fair play on the information highway, was caught napping at the wheel. However, governments are not interested in fair play—they want order.

The real problem with the Net—as far as the gatekeepers and policy makers are concerned—is that people are talking and publishing and exchanging information in an environment where traditional authority is essentially irrelevant. Littman's warning that, on-line, every opinion is as legitimate as every other opinion, is the clearest expression of the gatekeepers' fears. The danger is that, on the Internet, you don't have to have a Ph.D. to be taken seriously—but dangerous to whom?

The Net effectively circumvents traditional modes of authority. *Anyone* can be a pundit, and anyone can publish his or her own newspaper. Zapatista communiqués cannot be intercepted by the Mexican authorities before they cross the border to the North *because there is no border to cross*. Corporate media, looking at the North American Free Trade Agreement as an opportunity to exploit the free market with Mexico are unable to rely solely on official Mexican government statements when the other side's story is out and known on the Net. While the mainstream political discourse is circumscribed by the institutions of government, the ability to disseminate ideas and information through the mass media and authority founded largely on economic standards of status and prestige, the on-line discourse is open and interactive. Anyone can say whatever they want.

For people whose power is predicated on rigid social structures and political institutions, this is very dangerous. Traditional authorities whose power is predicated on neat, tidy geographical boundaries have no way of controlling the flow of information in the borderless Net—and in an information society, information is power itself. The Net isn't going to put Bill Clinton and Jean Chrétien out of business tomorrow—the on-line population still represents only a small fraction of the general population—but it provides us with a new, interactive model of political discourse. The danger is that it's training us to expect and demand the same level of interaction from traditional political institutions.

Faced with the threat of information anarchy, they respond by erecting barriers, marking some areas out-of-bounds and replicating geopolitical jurisdictional boundaries. The campaigns to regulate on-line content are not about pornography, hate propaganda or protecting the children. They are about power and restricting access to information. Offensive content and computer crime—little of which would be illegal in any other medium in the United States and Canada—are a pretext, the proverbial thin end of the wedge by which the gatekeepers can extend their control into an environment that they fear but don't fully understand. Few of us want to be seen as advocates of sleaze and indecency, fewer still would leap to the defence of a child pornographer or a Nazi, and that is just what the regulationists are counting on. By clamping down on unpopular content and behaviour in the name of law and order or the need to protect the vulnerable, they are establishing beach heads in the datastream. They fear the 'chaos and cacophony' that Judge Dalzell recognized, but by bringing order to it, the regulationists seek to tear the heart from the Net.

Efforts like the National Information Infrastructure Project and the Information Highway Advisory Council, which are official government policy in the United States and Canada, along with similar projects in most of the industrialized world, appear to encourage the information revolution. But this encouragement is not unconditional. The state's first priority is the preservation of its power and control over information. Injunctions against the export of cryptographic software, content controls and technologies like the clipper chip are individually insignificant, but together they reflect the campaign to exert civil power over the on-line polity. While publicly championing the information revolution, governments are in reality consumed with finding ways to establish themselves in an environment where borders are all but irrelevant.

CHAPTER SEVEN

All in a Day's Work

I. Road Warriors

ON A VISIT TO TORONTO a couple of years ago, I found myself killing time in the Eaton Centre food court. Food courts are grim places where the harsh throb of fluorescent light never fails to reveal the studied inanity of the urban shopping mall. It wasn't exactly my idea of a great place to hang out, but I wasn't in much of a position to complain. I'd arrived downtown uncharacteristically early for an appointment, and being a stranger to the city, I made a beeline to the only landmark I recognized north of Queen Street.

I had some work to do, so I woke up my laptop and started typing. Over the top of the screen, I could see a well-dressed businessman a few tables away, hammering away at the keyboard of an expensive black notebook and printing out impressive-looking reports on a portable colour printer. Here we were, two road warriors, with duelling laptops at 20 paces in one of the most desperately banal locations on the planet.

That's what the new office is all about in the age of information. Armed with the right hardware and a few key peripherals, information workers can set up shop just about anywhere. In the new office paradigm, distance, isolation and mobility are not barriers to productive labour, but merely other work environments. We can be as busy on a Formica tabletop—the kind with simulated woodgrain—as at the perfect ergonomic workstation... At least that's the official doctrine from the computer industry.

Portable and remote computing technologies evolved at a particularly furious pace, leaving us inveterate road warriors repeating "I remember when," like the refrain of some arcane incantation, even though when was less than a decade ago. I remember when armour-plated behemoths like the venerable Kaypro were considered portable; when mobile computing involved running an extension cord from the back of your station wagon.

Yet there was my fellow traveller—unaware that I was following him in rapt fascination—climbing the escalator to the mall's main floor so he could jack in his cellular modem and log onto his company's remote access server.

The remote workplace is where the leading edges of computer technology, software and telecommunications converge. In many ways, it's the office innovation of the last decade. The remote office, whether it's on the road or in a suburban home, exists outside of our culture's traditional understanding of workspace and time. In an economy that feeds on data rather than physical sweat and toil, *where* the work is done is essentially irrelevant. Information has no mass or volume and its circulation adds no expense to the original production cost. Thus, the information-based corporation sheds the trappings of location and geography. It assumes the raw character of a concentration of capital and data.

Freed from the imperative of location, information workers can conceivably operate from anywhere *except* the head office—assuming of course that there even is a head office. Telework and a mobile workforce are trumpeted as the great benefits of remote computing technology. They are expected to provide a hitherto unknown proximity to the market and services, quicker response time and greater workforce flexibility. However, in the rush to implement remote workforce strategies in information industries, pursuing the obscure goal of increased productivity at a lower cost, it's hard to shake the feeling that no one is looking too closely at the potential consequences for the economy and, above all, the worker.

I followed my fellow road warrior for a few minutes through the Eaton Centre's grand concourses. I watched as he precariously balanced his computer on a raised knee, entering figures into a personal finance program while he withdrew cash from an automatic teller machine. Upstairs, at the Disney Store, he called a shopping list up on his active matrix display, and asked the sales clerk to help locate gifts for his daughter.

The attractions of a fully digitized, mobile or geographical distributed workforce may be irresistible, but having the right technology doesn't mean we will ever eliminate office walls. The challenge is not to give everyone a laptop and send them home, but to understand how—and in which circumstances—telework and workforce mobility will improve the quality of work and working conditions.

II. No place like home

The office worker's daily commute has been a sign of status ever since the development of automobile suburbs following the Second World War. Every morning the highways and access roads to North America's major cities choke like a northern river in the spring thaw as hundreds of thousands of commuters battle rush hour traffic. It's a journey fraught with anxiety and dread, but also a badge of socio-economic status, a daily rite that affirms the prestige of those who, by virtue of their relative affluence, have been able to abandon congested city centres for the luxury of the suburbs. Hour-long gridlocks are a declaration of the binary opposition of home and work, of private and professional or occupational life, and consequently of the quality of that life.

The congested city is a creature of industry. It is a demographic manifestation of the industrial revolution's economy of scale. Mass production implicitly requires a mass labour force, so manufacturing centres like Detroit, Chicago, London and Manchester are as much vast concentrations of population as convenient loci of industrial technology.

However, for an increasing number of professionals here and elsewhere in North America the daily travail of getting to work is no more agonizing than carrying a cup of coffee into their home offices. They are on the leading edge of a trend that promises to irrevocably change the nature of the work environment.

The emergence of the personal computer as the essential tool of modern business and the corresponding boom in computer sales to the home market have given rise to two distinct kinds of home office workers. Contractors and freelancers have traditionally operated independently, but computers and subtle changes in the economy have allowed them to reduce operating costs and increase their flexibility by moving into the home.

On the other hand, major corporations have begun to explore the possibilities of telework, a system through which employees linked through advanced telecommunications and networking technology are able to work from almost any remote location. The traditional office is being joined by virtual offices defined by a network of teleworkers rather than geographical location.

The number of self-employed workers as a proportion of the labour force has been climbing steadily since the early 1980s, reversing a fifty year trend. The increase in outsourcing by companies seeking to reduce costs, and concurrent advances in information technology during this period, are largely responsible for the turnaround. The perceived advantage of outsourcing is that, with a large proportion of its workforce on contract and off-site, a company can reduce basic operating costs like real estate rentals and employee benefits. Indeed, the contract employee is the archetypal information worker, operating independently, and without the safety net of a benefits package from the relative isolation of a home office.

While home-based businesses and cottage industries were not uncommon before the information revolution, what sets the new home office apart is its economic significance. Home-based freelancers are doing the kind of work that used to be done in head offices, and much of the information and knowledge work that once required the support of the corporate data-processing infrastructure has been moved well beyond the corporate centre. For example, it can be hard to justify maintaining a full-time staff of analysts and planners to produce periodic reports if fully-equipped freelancers can do the work on a casual basis.

The related phenomena of outsourcing and downsizing—the former enables the latter—have helped to create the conditions for the growth of the micro-businesses that have become essential components of the information economy despite their size. Many of the most innovative and influential players in the new economy are refugees of the successive waves of layoffs and attrition that characterized big business in the 1980s and 1990s. Joseph Boyett and Henry Conn pointed out in *Workplace 2000* that as the information economy grows

> we will see the continued growth of truly entrepreneurial "micro-businesses"—those started up with less than four employees. Many of these will be started by executives or middle managers of large companies who are forced out as a result of cutbacks or downsizing. Being unable or unwilling to relocate to another large company, these displaced executives and managers will create their own small businesses using their own money, or money borrowed from friends or relatives.[1]

Micro-business growth fuelled the phenomenal boom in home computer sales that drove the information technology industry's expansion in the early-1990s, and sales of home-office systems have grown throughout the middle of the decade at an annual rate of 40 percent. The remote distribution of information-related work has only been possible with the penetration of computers in the home, and the technology, to a great extent, is what defines the home office. Basic organizational and administrative tasks—like filing, document management and typing—that would have required a full office staff, or at least a secretary, can be handled by a single professional with the right combination of software and computer hardware. The cost of setting up a competitive consulting operation in a home would have been prohibitive as recently as the early 1980s, but the development of personal computer technologies has provided information freelancers with the new economy's dominant means of production.

It goes almost without saying that the only *kind* of work that can be done from a home office is the production, manipulation or application of information. The typical contract worker in the information economy is a consultant, or specialist—the kind of skilled worker on which companies in an era of radical downsizing depend—and the home office is ideally suited to the realities of the freelance professional's work. Contract work simply doesn't provide the kind of financial security necessary to maintain a special work location, so combining residence with workspace is an attractive solution.

However, the relatively small scale of micro-businesses can be deceiving. While home-based information enterprises are typically one- or two-person operations they routinely emulate the strategy of downsized corporations by outsourcing part of their work in turn. The experience of Promève, a Montreal-based media-relations firm, is symptomatic of the freelance economy. The company is not so much a corporate entity as a loose network of professionals who are called into projects as their specialized skills are required. In effect, Promève is a clearinghouse of information skills that provides its participants with some of the security and profile of a traditional company without the overhead costs. The technology, and the freelance nature of the industry allow enormous flexibility, and Promève's client list includes major information-technology and food-industry players who, before the information revolution, would have maintained expensive media departments in-house.

As one Promève associate, media consultant Pascale LeBlanc, explained, "The size of a company doesn't matter in communications and services because all companies hire staff on specific contracts. We all have our information and our computers, and as long as you do the job and your clients can reach you, it doesn't matter where your office is."

What LeBlanc and millions of other freelance information workers have discovered is that, while there is strength in numbers, it doesn't have to come at the expense of independence and flexibility. Virtual organizations like Promève that exist only as long as required for a given project, are a definitive, though still far from the dominant, corporate structure in the information economy. Business, once conducted from centralized offices with large, full-time staffs, is undergoing a fundamental transformation. Work habits and business processes are changing as the firm gives way to the network. Nevertheless, it is equally true that most micro-businesses remain dependent on large corporations. Promève simply would not exist if companies like Claris and Apple didn't outsource all or part of their marketing projects and media strategy. Micro-businesses are creatures of the outsourcing phenomenon, but their proliferation and success have begun to have a corollary effect on their economic sponsors, the dominant forces in the information economy.

The flexibility of self-employed professionals like LeBlanc has attracted the attention of major corporations seeking to reduce operating costs and streamline their operations. Such cost-cutting measures dovetail with an increasingly significant trend to completely re-invent the corporate structure in the quest for greater responsiveness. Traditional corporate hierarchies, with their overheard and support superstructures, can be expensive to maintain and comparatively inflexible. In the de-regulated business environment of the neo-liberal global economy, and with technological developments changing the ground rules every day, such corporate structures are often at a competitive disadvantage. The corporation continues to exist as a concentration of capital and the central node of the decision-making network—but the network can be anywhere. Using the example of networked micro-businesses, where the partners are linked professionally and technologically without having to be in the same location, corporations hope to perform the miracle of becoming more competitive while reducing costs and increasing employee productivity and morale.

The idea that workers enjoy a better quality of life—and thus become better workers—when they are freed from the institutional restrictions of the traditional office environment has become a central management tenet the information economy. High-tech companies like Apple and Microsoft pioneered informal work environments in the 1980s, and the dominant theory today contends that, given the option of working from home, employees can maximize their personal and family time, with a corresponding increase in productivity. The equation is 'a happy employee is a more productive employee,' and what company doesn't want to have a more productive work force?

The solution is technological as much as it is a question of restructuring of workplace culture. Telework, where computerized employees participate in a company's daily operations from remote locations, is becoming increasingly widespread in the information economy, and industry analysts expect the number of companies offering their employees a telework option to rise sharply in coming years. Remote office work has moved into the mainstream information economy. More than a third of all North American households are equipped with a personal computer, and at least half of those are connected to modems, making telework at least a possibility for a sizeable chunk of the workforce. In Canada alone, sales of telework routers—hardware that allows workers to access a company's network from remote locations—are expected to rise from $4 million in 1995 to $24 million in 1997.

Many information industry executives believe that telework will quickly become the information economy's predominant method of company organization. According to Sid Livermore, the manager of IBM Canada's telework project, "the pressure for flexible work options is the result of companies seeking to be more competitive. It empowers employees to go in and do a job without delay... and everyone's looking to do everything faster."

Since 1992 IBM workers have been offered the opportunity to spend all or part of their work time away from the main office. The company provides a 'hotelling' solution—pools of shared desks and workstations—for Flexiplace participants who come into the main office, but a large proportion of employees could just as easily work from home. By the end of 1995 almost two-thirds of the company's employees could telecommute to work if a winter storm made the traditional commute impossible or just unpleasant.

III. On the road

The other side of the telework trend is the mobile professional, the travelling salesman's high-tech descendent. In their pursuit of flexibility and responsiveness, many companies have begun to dispense with offices altogether for many of their employees, sending them out on the road.

The travelling salesman is one of the business world's most romantic icons, driving down lonely highways, armed with a fistful of orders and a trunkload of catalogues and samples. He's the economy's front-line soldier, and though his goals haven't changed much in decades, the way he meets them has. Once a lone operator, the mobile professional relied entirely on his wits, only calling on the resources of the main office at scattered points throughout his route. The development of portable computing technology rivalling the power of desktop systems has allowed mobile professionals to bring their offices with them. As a result, companies whose workforce operates principally on the road have no reason to maintain the expense of a fully appointed office. With sales staff on the road for weeks or months at a time, many companies have begun to dispense with desks altogether.

Once arcane technologies like portable computers and wireless data communications devices allow mobile workers to interact with the company in real time. While a salesman used to come back to the office after a few weeks with a stack of orders, those orders now come in as he gets them, with the advantage that the company can count on a steady cash flow, without having to wait for the periodic crush of orders when salesmen return from the field. Conversely, retailers like the Gap have begun to rely on mobile representatives to maintain just-in-time inventories—instead of ordering for a whole season and risking the possibility of over-stocking, they buy for considerably shorter periods, thus reducing overhead and operating costs.

In the ideal telework model, the head office is less a physical location than a communications centre and an information clearing house. Before the information revolution workers had to work together in a specific location in order to share data, receive memos and shuffle paper. However business—particularly in information industries—is increasingly becoming less paper-bound. In theory at least, memos are as likely to be circulated by electronic mail as they are by paper, and the traditional binders of company data, customer and account information have been replaced by electronic databases. With employees communicating and accessing company records through their computers, it's a small step to replace desktop systems with laptops, and workers in many industries, from clothing salesmen to consultants, are transforming the office environment by taking it on the road.

But creating a mobile work force involves much more than equipping employees with cellular phones and laptop computers. Established business practices have to be adapted to take full advantage of the new technology. Moreover, after making the investment in computer hardware, peripherals, and wireless communications, workers have to be retrained if the investment is to pay off. Even in sales and service industries, where a large proportion of the work force has traditionally operated outside of the office environment, the new technology requires new skills. The ability to call on almost any information as they need it has made the mobile worker's job both easier and more complex than before. He has to be a polymath who doesn't just sell or consult, but analyzes inventory and stays informed on every aspect of his clients' accounts as well. Thus companies can become more flexible and quicker to respond to a client's needs by essentially being in two places at once, meeting one of telework's principal goals.

Convenience, flexibility, responsiveness and cost savings in real estate and facilities are one thing, but telework is only part of what has become a wide-ranging re-definition of the workplace. The so-called virtual office is an environment in which varied company functions are distributed among any number of remote locations. Without the structure implicit in the traditional office, the participants in the business process are supposed to be united by a sense of corporate common cause. They "won't just work for a company, they will be expected to 'join the team' and 'become part of the family.'"

Some of the people charged with the task of implementing this radical transformation have serious reservations. Julia Gluck, a partner in the Toronto-based consultancy Johnson and Smith International is concerned that "the lure of technology and the possibility of reducing costs may colour people's view of the virtual office. While you won't necessarily lose more than you gain in a virtual office, the possibility is certainly there. Man is a social animal. One of the consequences of the virtual office is that people will begin to feel isolated, and that will ultimately affect their quality of work. People learn most effectively through social interaction, and that can be lost in the virtual office."

It is impossible to overstate the importance of group interaction to the process of occupational learning and the creation of new ideas. The rapid pace of innovation characteristic of the information revolution has placed a heavy burden on workers in all fields to stay current with each succeeding generation of office technology. Combined with the imperative to learn the ropes and assimilate the new business practices implemented to take full advantage of computers, distributed work and the virtual office, the geographic isolation of individual workers becomes problematic. While Robert Logan asserts that "the intense social interaction and cooperation which surround computer use in the workplace parallel the experience in the classroom,"[2] such a fertile learning environment is predicated on the existence of a common workplace to start with.

The virtual office may ultimately succeed in undermining business' greatest asset—the corporate culture. Brainstorming and group-think are essential management activities. Group cohesion is particularly important in industries that produce the basic component of the information economy—ideas. The challenge is to find a means to replicate the sense of community and group interaction in a widely distributed environment. Corporate intranets and so-called groupware products, like Lotus Notes, are promising, but they only address part of the problem. Most computer-mediated interactions are asynchronous, allowing for high-volume exchanges of information but little opportunity for the cross-fertilization of ideas. Emerging technologies like real-time video conferencing promise to bring immediacy to the network, but only at enormous expense and with staggering hardware and bandwidth requirements. The simple truth is that the demands of the virtual office are still far behind the technology able to answer them. Only the largest companies can afford to implement comprehensive videoconferencing systems, and the expense and relative scarcity of high-bandwidth communications mean their use will be rare for some time to come.

The danger is that virtual office strategies will be implemented with little consideration of current technological limitations and possible consequences. Companies like Lotus and IBM have approached the issue with some caution, offering telework as an option rather than a requirement.[3] Despite the advantages inherent in distributed corporate structures, the path to the new workplace is fraught with peril... but more so for workers than for their employers.

IV. The machine bites back

This is one of the biggest—and most unpleasant—surprises of the information revolution: the computerized workplace can be hazardous to your health. Entirely new health care professions have emerged in the wake of the growing ubiquity of computers in most work routines. Only a few decades ago, there was no such thing as an ergonomist or industrial hygienist. Yet the occupational health consequences of the information revolution have made them as common as network administrators and information technology managers.

According to Edward Tenner, one of the incisive and insightful chroniclers of the effects of technology, repetitive strain injuries (also called cumulative trauma disorders) are the inevitable revenge effect of the information revolution. Carpal Tunnel Syndrome, one of the most common repetitive stress injuries, or RSIs, is caused by the thousands micro-movements required to use a computer keyboard. However, "the human being did not evolve to perform small, rapid, repeated motions for hours on end,"[4] and the most basic activity in the modern workplace is inherently dangerous.

RSIs come in a wide variety of forms and have many causes. They are all characterized by an inflammation of muscles or tendons, which may pinch nerves and cause numbness and pain. While almost anyone who has typed at a keyboard or pushed a mouse around a desk for a few hours may start to feel some discomfort, a full-blown RSI can be seriously debilitating, and they account for more than 60 percent of all occupational injuries in the United States. Ten times as many new cases were reported in 1993 as in 1982.[5] It's not clear whether this increase is due to the widespread use of computers (Tenner is quick to point out that RSIs are still more common in traditional industrial occupations) or to increased reporting. However, many industrial hygienists and ergonomists believe we have only seen the beginning of what will soon become a serious epidemic of computer-related injuries.

At Concordia University in Montreal industrial hygienist Karen Ward evaluates and controls health hazards in the university's work environment. Her job is to prevent occupational health problems before they happen, and try to make sure they don't repeat when they do occur. She doesn't like what she sees in the modern workplace. "We have an aging work force, and people have been working on computers in large numbers for ten to fifteen years. That is the latency period for most repetitive stress injuries, and I'm certain we're going to see a major outbreak of carpal tunnel syndrome fairly soon."

Most RSIs are the result of untrained, and unhealthy work patterns. Aside from the micro-movements necessarily to enter data on a computer keyboard, we routinely force the muscles of our hands and arms to repeat unnatural and dangerous movements, hyper-extending our wrists, twisting our hands to one side or the other over and over again all day long. Simply typing—the activity at the very heart of everyday computing—can be begging for trouble. The keyboard as we know it today has been in use for about a century, but keyboard-related RSIs seem to be fairly recent phenomena. According to Ward, the reason why generations of typists, slaving away on mechanical Remingtons and Underwoods, didn't fall victim to carpal tunnel syndrome or similar ailments is that, unlike modern computer users, they knew how to use their tools. Computer users typically pick up a keyboard with little or no training, but "most of the people who used keyboards in the past were professional typists. They were trained in proper keyboard technique, but most of us aren't taught that anymore. And the principal cause of RSIs is poor technique."

Technological limitations forced typists to break up their writing routines with pauses and micro-breaks every time they hit the carriage return or put in a new piece of paper. As Tenner points out, the mechanical typewriters of the past were, by accident, simply designed more ergonomically than the flat slab keyboards common today.

> Manual keys literally had a spring to them that prevented the sharp impacts that firm strokes on a computer keyboard can bring. (Try this comparison with your computer and an old manual machine.) The mechanics of manual keys also encouraged straight wrists, making less likely the arched and suspended hand positions that seem to be risk factors for CTS.[6]

Moreover, manual typewriters' tendency to jam at high speeds forced typists to maintain a relatively slow pace. Even electric machines like the IBM Selectric which overcame this problem, forced typists to pace themselves by requiring frequent page changes and pauses to white-out errors.

The pace of data entry today is symptomatic of what may turn out to be the biggest health hazard in the modern workplace. Computers operate much faster than human operators are capable of keeping up with. Huge volumes of data are available at the speed of light over company networks and the Internet, and the growth of a global information economy means that a North American business in constant touch with markets in Europe and Asia has to be in operation around the clock. The forty-hour-a-week, nine-to-five work schedule may have suited the pre-computer age, but in the new workplace sixty, even eighty-hour weeks are not only common, they're quickly becoming the norm.[7] Moreover, the very technology that makes this work rhythm possible has been turned around to make sure that workers don't fall behind. As early as 1985, growing numbers of companies began implementing computerized employee monitoring systems to keep track of workers' office performance.[8] Software designed to count keystrokes and catalogue employees' pauses and breaks has created a sense of siege in the information workplace. Employers can intercept electronic mail messages at will, allowing them to keep track of workers' plans and guage employee morale. One colleague who was planning a career change once cautioned me to keep my e-mail to his office address innocuous. He knew his mail was being read.

This kind of work environment can't be anything *but* stressful, and while, at such an early stage in the information revolution, it's hard to say how it will affect employee productivity and economic performance, it's nonetheless clear that there will be a price to pay.

> At executive seminars run by Harvard Medical School, technology-induced exhaustion is now the most popular session. The combination of new tools and the new fast-paced global business environment has, for many, created a new tempo and rhythm for work. For many, we've gone from Mozart to M.C. Hammer.[9]

You need only look at the legions of laptop-equipped professionals in any city park on a sunny summer day to see how bad it can get. The metronomic pulse of work in the information economy doesn't even stop for the brief customary period of rest and recuperation at the mid-day meal. Instead of taking a breath, watching the birds light on statues and making slow work of a sandwich, they wolf down their lunches without shifting their gaze from the computer screen for a second. Considering that many of our best ideas come to us over lunch (the others arrive just before falling asleep or in the middle of dinner), it's hard not to imagine disasterous consequences for the economy.

V. Division of Labour

The perils of the new workplace pale in comparison to the socio-economic dislocations that follow in its wake. Information has become our economy's dominant commodity, and a majority of workers use information technologies in their daily work. However, those workers who are *directly* involved in the production and exchange of information remain a relative minority and it's becoming clear that anyone who doesn't have the skills to operate the means of production will soon be relegated to the menial fringes of the economy. The situation is somewhat analogous to the industrial revolution of the late-eighteenth and early-nineteenth century, when the power of human and animal muscle, sinew and sweat was replaced by coal and steam as the engine of the economy. Whole trades and cottage industries were eliminated when the great capitalist barons discovered that a child at a steam-powered loom could out-produce an experienced traditional weaver, or that a hundred highly-paid iron founders could easily be replaced by assembly lines.

It goes without saying that only those who master the necessary skills will participate in the new workplace. However, due to the complexity of the means of production in the information economy, those workers whose jobs were eliminated in the last wave of downsizing and lay-offs may not have the option of reinventing themselves as information workers. This is hardly a radical assertion. In *The Fifth Language*, Robert Logan observed what may become the ultimate challenge facing individual workers in a growing and increasingly competitive labour market.

> Just as literacy created a new social class and a new form of privilege and economic opportunity, the use of computers may do the same. Those who will be able to exploit the power of computers will have an important advantage over those who cannot, just as those who were literate had an advantage over those who were not.[10]

The stark prospect for workers without these skills is that they must acquire them or remain trapped in occupations tied to industrial and not information production. The problem is that industrial production and manufacturing in the information economy is itself being revolutionized by computers. Any form of production that involves the application of information, from automobile assembly to mining, can be automated to some extent, and in the pursuit of lower operating costs and higher production output, automation has become the central theme of North American industrial development—at the cost of workers' jobs.

Those fortunate few who retain their positions will have to acquire new skills, but it has already become clear that the demand for workers trained in information technology is far greater than the supply. Companies simply aren't retraining their work force to accommodate this demand, and secondary and post-secondary educational institutions are producing graduates unprepared for a high-technology economy. Scarcity allows those workers with the appropriate skills to take advantage of the demand, receive higher salaries, and exert a disproportionate influence in the labour market, with the result that the labour of traditional industrial workers inevitably will be devalued as the wages and power of the new labour aristocracy rise—industry has a finite pool of capital to allocate to wages.

A new labour aristocracy is implicit in the promise of the information revolution. For example, the wages of computer programmers, the essential workers in the most lucrative segment of the information economy, rose at nearly twice the rate of *all workers* between 1990 and 1994. Moreover, jobs involving computer use *in any capacity* are more highly valued and more lucrative than other occupations. Computerized workers earn as much as 15 percent more than their non-computerized colleagues.[11] The discrepancy is a recipe for crisis. What we are witnessing in the information economy is not the creation of new wealth, but a new socio-economic stratification among workers. Wired workers will make more, and unplugged workers will make less, possibly leading to the creation of an underclass of technological have-nots who see the promise of the information revolution collide with their declining wages. The irony is that computers and automation require fewer workers to perform the same tasks, so the ranks of have-nots will probably grow, thus reducing the size of the market for information products. And that's only assuming that the big players in the information economy continue to employ North American workers at all.

The emerging division of labour is also reflected in the geographical distribution of economic development. The modern, congested city is a creation of the industrial revolution. In the manufacturing economy before the information revolution, economies of scale were principal factors in a company's success. Large population concentrations, and thus significant pools of available labour, were essential in the development of heavy industries. However, information industries are considerably less labour-intensive, and both the means of production and the product itself are portable. With telework and the mobile office, the leading corporations of the information economy simply aren't obliged to set up in urban areas—or in any specific place at all, for that matter—and many have already fled the congestion and crime of the cities.

It's true that high-technology industries tend to sprout in clusters, in the fertile soil around universities and research laboratories, where they can cross-pollinate with similar companies and feed on the talents of young engineering and computer science graduates. Silicon Valley developed as an annex to Stanford University in the suburbs of San Francisco. The companies that line Route 128 outside of Boston were nurtured by the nearby Massachusetts Institute of Technology. They needed the support of big city universities and access to venture capital in downtown financial districts. Once mature, however, most relocate to locations where land and labour are cheaper and taxes are lower.[12] A good chunk of the information economy has found a home on the range in the American Southwest

But it doesn't stop there. The information economy is a creature of neo-liberalism, an ideology with free trade as its key doctrine. With the relaxation of trade barriers, computer manufacturers have moved increasingly large parts of their operation to Mexico and Asia. Proudly American companies like IBM and Apple have their systems built and assembled by cheap labour in Malaysia, Singapore and Thailand, and the production of the silicon chips that hold the memory of these computers has been largely shifted to plants in Japan and Korea. Other information-based companies are following the exodus.

Management, planning, research and development and software design remain largely North American activities, but conceivably even they could be transferred overseas. After all, a highly-trained computer programmer in India—where the average income is far lower—will probably be willing to work for far less than his American counterpart is earning. More to the point, an economy based principally on management and development will provide few opportunities for those industrial workers already fighting for dwindling positions. It's not that North America will ever be industry-free, but a significant trend in the information economy is a desire to see how far it can go before the economy collapses.

Corporate leaders have lost their long-term perspective in the rush to apply the tools of high-technology to every part of the economy. It is not possible—or even desirable—to roll back the information revolution, but we have to restructure the economy so all levels of industry, from coal miners to software developers, support each other. Indeed, one of the great fallacies of neo-liberalism is that free trade necessarily benefits all concerned. The dismantling of trade barriers, and the migration of capital, industry and jobs enabled by information technology might ultimately have disasterous effects. For it raises the old question—who will buy the economy's products when all of the jobs have moved away? Workers will certainly have to acquire new skills and accept the transience and insecurity of a freelance and geographically fluid economy as a fact of life.

CHAPTER EIGHT

In the News

I. Open Secrets

THROUGHOUT THE SUMMER of 1993 an uneasy pall of silence hung over the Canadian news media. Lurid stories had been circulating around restaurant tables and office water coolers since the arrest that January of Paul Bernardo for the brutal rape and murder of St. Catharines, Ontario teenagers Kristen French and Leslie Mahaffy. The trial of Bernardo's estranged wife and accomplice Karla Homolka was expected to be the place where the truth about the horrific crimes would finally come out, but it didn't happen that way.

On July 6, Justice Francis Kovacs, presiding at Homolka's trial, clamped down on the proceedings with an almost unprecedented publications ban. Judge Kovacs's decision was supposed to protect the integrity of the trial process and ensure that Bernardo would have an unprejudiced jury in his trial on first-degree murder charges, to be held the following year.[1] The courtroom was closed to all but 88 people, including the defendant, the lawyers, the victims' families and members of the Canadian media—American journalists, who would operate beyond the reach of Canadian law, were barred. However, Canadian news media were forbidden, under threat of a contempt of court citation, to publish any details concerning the evidence presented against Homolka, or her confession and plea until Bernardo's trial to ensure that his jury would not be tainted by knowledge of the gory details. Customs agents seized American newspapers containing trial coverage at the border, and cable television networks, under government pressure, cut the signal from U.S. television stations when *A Current Affair* and its clones broadcast reports on the case.

However, something had begun to happen on the Net. Unsubstantiated reports about the crimes for which Homolka was sentenced to two concurrent twelve-year terms had begun to trickle into a handful of Usenet newsgroups, and by late summer, the details of her confession had become a wide-open secret. On two newsgroups—ont.general, and a special forum mobidly entitled alt.fan.karla-homolka—information concerning the St. Catharine's murders was presented in a mix of plausible hearsay and outlandish rumour... but much of it, gathered in an FAQ (a "Frequently Asked Questions" document) at the end of August, was consistent with the few facts in general circulation.

The FAQ itemized the atrocities committed by Bernardo and Homolka. It reported that the victims were held, repeatedly raped and tortured over a period of several days to weeks, they had been hobbled by their abductors, who cut tendons in their legs with surgical instruments stolen from the veterinary clinic where Homolka had worked. There were charges that Homolka and Bernardo documented their crimes on videos, and that one of these tapes apparently showed the rape of Homolka's unconscious 14-year-old sister Tammy, whose death was originally ruled accidental. Despite errors in some of the details the FAQ turns out to have been astonishingly accurate in its broad account of the crimes. The managing editor of the St. Catharine's *Standard*, one of the few journalists in the courtroom, was amazed at how accurate the local gossip was, and when it hit the Net, it became global.

A university student named Neal Parsons became something of a Net celebrity, proudly accepting the moniker Neal the Trial Ban Breaker. "It was originally begun on a lark," he explained. "Someone else created the alt.fan.karla-homolka newsgroup, and they spent the first two weeks discussing the legal ramifications if anyone said anything. By happenstance, I was having lunch, and my source told me a brief vignette about the trial and I decided to post and see what happened."

Before long, the volume of trial information on Usenet and on electronic bulletin board systems grew. Unlike Parsons, who posted information under his real name, many users hid behind pseudonyms and phony e-mail addresses. Even if the Canadian authorities had wanted to track down violations of the press ban, most of the flood of information was beyond their ability to control. Rod Macdonnell, a veteran investigative journalist for the Montreal *Gazette* observed at the time that the Internet "has robbed the courts of their teeth. You can't punish offenders if you can't find out who the offenders are."

While publication of the details of the trial was forbidden, there was no injunction against casual conversation and gossip. Judge Kovacs evidently hadn't considered that casual conversations on the Internet can involve several million people worldwide. Like most of the current generation of jurists and legislators, he was utterly unaware of the power and flexibility of the new medium. Indeed, the speed at which information can be distributed to the millions of users on the Net makes sanctions like the publications ban almost irrelevant. Judge Kovacs' order is an example of the kind of legal thought belonging to an earlier, technologically simpler age, but it also demonstrated how dangerous such ignorance can be. The rapid spread of rumour and gossip on the Net certainly undermined the publication ban's intent. With no editors, and no way to check sources, bizarre stories, like one about Homolka masturbating with either a victim's hand or head (the informant wasn't clear), spread wildly along with more plausible hearsay. By the time Bernardo finally stood trial in Toronto in the spring of 1995, revelations of the defendant's brutality and depravity—as horrific as they were—seemed almost pale next to the lurid gossip that most Canadians had accepted as truth in the absence of news coverage for more than a year. "When we were arguing against the ban, we said that, if we published what happened, and what was said in the trial, it would be right, or at least correctable," St. Catharines *Standard* Editor Murray Thompson explained. "Compare that to the rumour mill where it's not correctable, and in no way accountable. And that's exactly what happened."

For all its shortcomings, the alternative coverage of the Homolka trial reflected the emergence of an entirely new and interactive news medium. Parsons circumvented the press ban because he could, or more precisely, because the technology gave him the means to reach a potential audience of millions. Using the dominant Internet technology of the time—by 1994, the Web was only just gaining popularity among Net users—he published what was, in effect, a digital simulacrum of a newspaper. In many respects, the alt.fan.karla-homolka FAQ is indistinguishable from the thin broadsheets published by independent journeyman printers with aprons full of lead type in the eighteenth and early-nineteenth centuries. What it lacked in editorial accountability and breadth of coverage, it made up for by being regularly updated and, arguably at least, by serving the public interest. Justice must be *seen* to be done, and in the vacuum created by the press ban, Parsons' efforts were the only source for that information. The Net holds no secrets.

II. Black and White

Even before the Internet and the information superhighway entered the public lexicon, the mainstream print media pursued an obsession with electronic news delivery systems. In the late 1970s, American and Canadian newspapers began trials of technologies that, it was hoped, would lead the news industry into the twenty-first century. Television was seen as the newspaper industry's biggest threat, and it was believed at the time that the print media's future lay, at least in part, on the small screen. In fact, the proposed electronic delivery systems were abject failures. They were generally slow, inflexible and difficult to read on the low-resolution text-only displays of the day.

Canadian videotext services—proprietary electronic text retrieval systems—like Telidon and Alex wilted from lack of interest in the 1980s. Other services, like Infomart, a joint venture between the Southam and Torstar newspaper chains in Canada, Knight-Ridder's Dialog and Dow Jones News Retrieval had a measure of success among business and professional users, and they demonstrated that at least some users would be willing to pay for news in an electronic form. However, their great expense kept these services out of the reach of most potential consumers, and are more analogous to archives and clipping services than daily newspapers. Teletext—the airing of static text on specialized television channels—promised to bring electronic news delivery to a mass audience, but while it continues to enjoy a marginal existence as a community announcements bulletin board on most North American cable systems, it has been all but abandoned by the newspaper industry.

The emergence of the Internet as a mass medium in the early-1990s gave new life to the dream of the electronic newspaper. The relatively low entry requirements that allowed alternative news providers to tap into the vast potential market of Internet users made the Net that much more attractive to newspapers. After all, every step in the production of modern newspapers up to distribution is almost completely computerized. News is just another information product that can be packaged and distributed on-line relatively easily, and it has become plainly obvious that news and information services are the principal goal of most users' Internet voyages. All that the newspaper industry needed was a single hot story to attract them. Almost on cue, it got one in one of the biggest media circuses of the century.

The Internet loves a good story. When you get 30 million people in the same space, the human desire to gossip becomes overwhelming. The Net is the world's biggest virtual water cooler, and the mundane daily activity of trading opinions, information and rumour is merely magnified by the size of the crowd and amplified by the speed of transmission. The 1995 trial of O.J. Simpson for the savage murders of Nicole Brown Simpson and Ronald Goldman—in which he was ultimately acquitted—established Internet users' insatiable appetite for news. After the major network trial updates, CNN's football-style play-by-play coverage, and the encyclopedic exposés in *Time*, *Newsweek,* and *People* magazines, it was surprising to see the torrents of O.J. information in the datastream. Yet there it was, from the alt.fan.oj-simpson, alt.fan.oj-simpson.drive.faster, to alt-fan.oj.simpson.gas-chamber Usenet newsgroups. Users from the U. S., Canada, Belgium, Japan and elsewhere debated *whether* he was guilty, *why* he was guilty, and why death by cyanide gas would be too good for him if he *was* found guilty. There were users whose whole lives—judging by the volume of their postings—were devoted to proving Simpson innocent, or to arguing that Nicole Brown Simpson was stabbed so violently that her head was almost severed from her body *because she deserved it.*

It didn't take long for newspaper and magazine publishers to take advantage of this enormous potential market. *Time* Magazine and the *San Francisco Examiner,* among many others, used the World Wide Web to bring that much more depth to the on-line O.J. coverage, and demonstrated, for the first time, how potent the Net could be as a supplement to traditional print media. *Time*'s and the *Examiner*'s trial coverage sites were easily the best information sources for those of us with nothing better to do than follow the minutiae of the trial. Users could read the latest trial transcripts, view photos of evidence and download Simpson-related news stories and features. The sheer quantity and depth of the information was astounding, and far exceeded the amount that could have been published in a daily newspaper or weekly newsmagazine. *Time* even sought the involvement of the readers in a special interactive forum. One user posted a message titled *OJ OJ OJ OJ OJ OJ OJ OJ OJ !!!! I'm sick of hearing about it!* but it was clear that most people weren't.

Once the audience had been established, it was only a matter of time before publishers realized the medium could be made to pay for itself, and it was during this time that advertisements began to appear on newspaper Web sites in any volume. Internet advertising typically takes the form of non-intrusive graphical links to a sponsor's home page. This kind of banner advertising, pioneered by *HotWired*, the on-line version of *Wired* magazine, offers several advantages over traditional print advertisements. By clicking on the ad, users can be transported directly to the sponsor, where they can be exposed the full force of a multimedia pitch, try out the latest product, or place an order by e-cash or credit card. Advertisers can use this interactivity to keep track of who follows what link from where, letting them refine their advertising strategies with a precision impossible in traditional media.

At least that's the theory. In practice, however, banner advertising has been disappointing. The total revenues from Internet advertising in the U.S. didn't top $50 million in 1995. Compared to the revenues from television—$60 billion—and radio —$10 billion—selling advertising on the Web is a pathetically small business.[2] It's not surprising. The Net may be growing exponentially, but it remains a young medium. On-line advertising was almost unknown as recently as 1994, and despite the fervent desires of managers and executives that the information revolution will open broad vistas of revenue opportunities, it will doubtless be some time before the Internet has penetrated far enough to be able to pay for itself through advertising. Moreover, the application of traditional advertising techniques to an entirely new medium is doomed to fail. The model of on-line information consumption is radically different than reading a magazine or newspaper, so it stands to reason that revenue models will have to be just as different.

In view of the apparent failure—at least in the short term—of on-line advertising, publishers have begun to take a second look at the possibility of charging subscription fees. When the Internet first emerged as a mass medium, the conventional wisdom held that it would be impossible to get users to pay for on-line information services. Charging a fee for access to information went completely against the hacker values of the Net's pioneers, after all. The argument went that, with so much *free* information available for the taking, users would simply choose not to patronize a service that charged a fee. Besides, concerns about security would probably discourage most people from using their credit cards on-line.

In the three short years since the introduction of NCSA Mosaic, everything has changed. The Net's phenomenal growth has assured that the pioneers who lived by the traditional on-line values of free speech and free access have been overwhelmed by ordinary users perfectly willing to pay for on-line services in the same way that they would pay for any other product in any other environment. The question of credit card security has become something of a non-issue with the development of secure browser and server software, and an unprecedented agreement between Visa, MasterCard, Microsoft and Netscape to develop the industry-wide Secure Electronic Transactions standard. The newest users are prepared to pay for information and services, and they have the means to do it easily and with a considerable measure of security. What they need is a product that they will be *willing* to buy.

The obvious solution was to open videotext-style information services like Infomart to a general market through Internet gateways, and though there remains a great deal of interest in this approach, the big news in the summer of 1996 was the debut of on-line versions of the *New York Times* and the *Wall Street Journal*. Both trade on their prestige by charging users an annual subscription fee for access to full text articles and archived back issues. It remains to be seen if these services will generate the kind of revenues their publishers hope for, but it's clear that the subscription model for on-line newspapers can only be applied to publications of the stature of the *Times* and *Journal*. *USA Today* abandoned its subscription fee plan shortly after implementing it in 1995, and in most respects, on-line newspapers are better suited as archival services than as sources of daily news.

The principal obstacle to the wide consumer use of daily newspapers—as newspapers—on the Internet is the interface. Fanciful predictions of the 'newspaper of the future' in which each day's news would be selected by intelligent agents and downloaded to some kind of electronic reader or browser contradicts the newspaper's traditional social role as a cheap, timely and portable medium providing a broad spectrum of information. Nicholas Negroponte, for example, muses that the newspaper is simply an "interface to news" that could now easily be electronically transformed into an interface that provides news on demand,[3] but Negroponte's optimistic vision of news tailored to specific tastes ignores the fact that the newspaper traditionally transcends individual preferences. News is general by definition, and newspapers do not provide readers only with what they *want* to know—though many, out of preference, might turn to the sports or comics pages first—but with a wider context and the unexpected.

Historically, newspapers have tried, and largely succeeded, to create a consistent interface for the widest possible range of information. The familiar package allowed users to consume the disparate information in a ritual manner, a "publication readers would come to expect—creating, in other words, a habit."[4] That habit arrives on the doorstep every morning in a form that can be read at the breakfast table, in the bath or on the bus. It can be tossed out after use, clipped, given to a friend or coworker, or used to wrap fish and chips. Most importantly, while Negroponte's agents require electricity, apparatus and processing power—and will thus come at a significant cost for the foreseeable future—newsprint is cheap enough for even the poorest readers to use.

Moreover, the electronic newspaper as a mass medium is a classic example of re-inventing the wheel. Aside from the ability to randomly search live text—only an advantage when accessing archival materials—electronic delivery systems don't offer any advantages over printed newspapers. Most of the information on-line is still in text, and the essential skill of the Internet user remains his literacy. The information revolution itself has created something of a revolution in *print* publishing as, every week, publishers come out with thousands of titles explaining what the Net is, how to use it, and what it all ultimately means. Black ink on white paper remains the most efficient and effective means for reading text. Of all our traditional news media, daily newspapers will likely remain the most resistant to the changes wrought by the information revolution. After decades of declining readership, newspaper circulations have leveled off since access to the Internet became a practical reality for the vast majority of information consumers, actually rising in 1995.[5] Significantly, the newspaper industry continues to claim a larger share of money spent on advertising than any other medium.[6]

III. The Digital Academy

The challenges facing other print media are somewhat more complex. The vitality of the traditional daily newspaper is based on the simple fact that no other medium is suited to its social function of providing the public with timely, broad information and context in an extremely inexpensive package. On the other hand, paper-based publications that serve limited constituencies or whose principal purpose is to act as an archive can't compete with electronic delivery.

These are likely to be subverted by the information revolution.

> As archives for information and knowledge, they have long been surpassed by electronic media, and with the Internet they have also been surpassed as distribution vehicles. Rather than printing all its articles on paper and mailing them to their subscribers, journals could distribute their material electronically...[7]

Electronic publication may ultimately prove to be the salvation of academic journals and related periodicals, many of which have been forced to fold in recent years due to increasing competition, production and distribution costs in a shrinking market. In 1995 American and Canadian research libraries spent 93 percent more to acquire four percent fewer titles than they did a decade earlier.[8]

Scholarly journals exist to provide a forum for the academic peer review process, and as an archive of past and current thought, thus the question of whether or not electronic delivery is suitable for a mass market is moot. The goal is to provide a narrow market with easy access to journal contents, and the Internet may allow publishers to do just that at a relatively low cost. However, it isn't quite clear how subscribers will pay for the service. One solution that has received a great deal of attention in on-line academic circles would see journals funded entirely at the publishing phase, with funds that would otherwise have been used to pay for subscriptions. With this model, libraries would not actually have to pay for individual publications. While the Net has not yet replaced traditionally published and distributed journals, many have found their way into the global datastream as electronic supplements to the print versions. Nevertheless, academic periodicals will certainly continue to face the challenge of survival as education funding is increasingly cut back. Preservation of their essential role will depend on the willingness of institutions and governments to subsidize journals' migration to the Net, with the ultimate goal of reducing the *total* acquisition costs.

The information revolution is likely to have little negative impact on most print media. Daily newspapers remain fairly secure by virtue of the superiority of print technology as an inexpensive mass news medium. More specialized publications, already threatened by economic circumstance, may actually benefit from a migration to electronic delivery systems, and in any event, studies have shown that the vast majority of Internet users see on-line publications as a supplement to, rather than a replacement for their print versions.[9] However, the alternative press has found itself in a no-win situation. Most alternative newspapers have begun their digital hejira only to find that the Net can only subvert their social function. In effect, they have nowhere to go when the constituencies they represent abandon traditional media for new technologies.

IV. The Alternative Press

The technological forces undercutting the alternative press have been accelerating for several years. Listservs, developed as a medium for academic discussion in the Internet's early days, provided alternative groups and non-governmental organizations like the Institute for Global Communications with the technology to distribute digital newsletters to large lists of subscribers. Other alternative publications, like the hacker journal *Phrack*, were available principally as downloadable text files or e-zines on BBSes and FTP archives. However, the development of World Wide Web browsers like NCSA Mosaic and Netscape Navigator brought a measure of maturity to the burgeoning on-line alternative press. The political and cultural groups who flocked to the Web hoping to establish some kind of virtual community presence soon discovered that they could use the medium to spread the word about their respective causes without having to rely on the often indifferent mainstream media. More importantly, while listservs were closed media, requiring readers to subscribe, Web pages are public; in theory at least, potential readers will simply happen along to a publication linked to a related site.

In most respects the coverage of the alternative on-line press overlaps the principal function of alternative newspapers like *The Village Voice* in New York City and *The Mirror* in Montreal. Unlike dailies, which serve the public interest by neatly packaging current events as objective journalism (they also serve their publishers' interests by making money, but that's another issue), the paper-based alternative press is the traditional vehicle for what has come to be known as advocacy journalism—news coverage explicitly in the service of a particular cause. The role of the alternative newspaper is to provide a forum for a wide spectrum of activist and interest groups to air their grievances and expose the injustices that they believe are committed against them.

However, these groups' varied missions are rarely complementary. You need only scan the pages of an alternative newspaper to see the often bitter competition for editorial space that largely defines the medium. In Montreal's *Hour* (a newspaper to which I contributed a regular column and numerous feature articles over a period of several years) news items were grouped under rubrics that didn't indicate beats or issues so much as describe their respective constituencies. Significantly, the gay issues correspondent was gay, the reporter who covered the refugee beat is an immigrant from the developing world, and the journalist on the women's issues beat was a woman. While such beat assignments certainly serve a practical purpose—a gay reporter will have a greater knowledge of the gay community and issues than would a straight colleague, for example—it also underlines the advocacy function of the alternative newspaper. It is disturbingly common in the alternative press for members of distinct communities and groups to be assigned to cover those groups *to the exclusion* of anything else. While many are gifted journalists, often much more tenacious and incisive than their mainstream counterparts, their implicit role is as the representative of their individual communities.

The alternative newspaper's secondary function as a review and schedule of artistic events is far more straightforward. Indeed, the vast majority of readers, many of whom do not belong to one of the constituencies represented in the news coverage, pick up these publications week after week simply to find out what act or production is playing where. The listings section is, in fact, one of the alternative newspaper's defining characteristics, and arts and entertainment advertising typically accounts for the majority of its revenue. Mainstream dailies and local television news programs have, in recent years, begun to erode their monopoly on alternative arts coverage. On one memorable occasion, almost every media outlet in Montreal from suburban community newspapers, to the dailies, local television and the three alternative weeklies provided blanket coverage of the city's Fringe Festival, indicating that 'fringe' may not be synonymous with 'alternative' after all. Despite their efforts, the mainstream media have yet to match the quantity and breadth of the alternatives' arts listings.

Nevertheless, the alternative press is in trouble. The alliance of disparate interests that has characterized its mission was always uneasy at best, and was originally formed because any one group simply couldn't afford the cost of producing, printing and distributing a newspaper, to say nothing of overhead expenses like office space and sales staff. By combining coverage, idealistic activists and canny entrepreneurs alike could serve disenfranchised communities and, incidentally, translate their loyalty and support into advertising revenues. The functions of the alternative press are being usurped by thousands of independent on-line publications springing up all over the Web. The costs involved in setting up a Web page are minimal and the potential audience is immense, making the imperative of ambivalent alliances considerably less urgent. The gay community can get its message out without having to share space with punk rockers, and the alternative arts scene can represent itself without appearing next to advertisements for phone sex. In effect, the emergence of alternative news media on the Internet has made publications like *The Village Voice* and *Hour* all but irrelevant.

Even the arts listings aren't safe. The paper press is at a distinct disadvantage when it comes to organizing and accessing large quantities of information. Anyone who has ever thumbed through the dozens of pages of theatre or pop music schedules, and lists of bars and galleries that appear at the back of every alternative newspaper would likely agree that printed lists are an inadequate medium for this information. However, the Internet provides sophisticated tools for searching huge databases for specific information. Even on-line directories like Yahoo! and Excite offer quicker and easier access to a far wider variety of data than a newspaper publisher could even think possible. The principal value of the alternatives' art listings is their local focus, but even here, there is no reason why an enterprising Web developer couldn't set up a database for local artists and entertainers, and at a far lower cost.

On-line arts information databases have only begun to appear in appreciable numbers on the Internet. However, individual theaters, galleries and pop music groups have already established a significant on-line presence, offering listings that are both far more detailed and easy to use than anything in an alternative newspaper. But the most important development in this area—possibly the technological coup de grâce—is the entry of Microsoft into the cultural listings business with its *Sidewalk* service. In development at the beginning of 1997, there will be *Sidewalks* for every major city in North America, inter-connected and with the full, multi-billion dollar resources of the world's biggest software company behind them. Significantly, Microsoft has been luring staff away from the alternative weeklies to develop content for the service, including almost half the editorial staff of Montreal's *Hour* magazine!

To make matters worse, the size of publications like *The Village Voice* and *The Mirror* has been dropping since the early-1990s, and even free distribution—allowing publishers to print 100,000 copies of a newspaper and claim higher circulation—doesn't seem likely to stop the erosion. Many of these publications have already found their way onto the Web. *The Voice*, for example, maintains an excellent site carrying the full text of current issues and an archive of past articles, but it seems unlikely that simply pasting a printed newspaper into a digital environment will change the equation. As Negroponte correctly observed in *Being Digital*, the paradigm of information consumption on the Internet is radically different than in print media—consumers *pull* information rather than having it *pushed* at them, making the business of information "more of a boutique business."[10]

This is perhaps only true of the alternative press, which has traditionally provided a forum for diverse opinions and perspectives in much the same way that Neimann Marcus or Ogilvy's house independent boutiques under one roof. However, the Internet has the potential to encompass all of these interests in the same data space, subverting the *need* for the alternative newspaper. Whether *The Voice, The Mirror* and similar publications survive the information revolution ultimately depends on whether they can redefine their mission in the information age.

V. Pictures at Eleven

For most of its history, television has been tweaked, bent and twisted in apparently vain attempts to make it more than it really is. The endless pilot projects and plans in the 1980s and 1990s that were supposed to bring interactivity to television without altering the information landscape missed the point entirely and were, by definition, doomed to failure. On one hand, the proposed interactions—alternate endings to television programs, optional camera angles at sporting events and video-on demand—simply cooled an already cool medium. On the other hand, none of the projects—Videotron Inc.'s Videoway service in Canada is a prime example—has addressed the question "interactive with what?" Such as it was, interactive television offered greater viewer options without providing more information. Significantly the same cable television networks that were its biggest boosters five years ago have set up their own Internet access service and have begun to preach something called convergence.

The doctrine of convergence has acquired the status of universal truth in government and the television and computer industries. The so-called information superhighway policies of the American and Canadian governments are predicated on the doctrine, promising information access on one hand, and a 500-channel television universe on the other. Dogmatic convergence proponents like Negroponte imagine a rosy future when all of the functions of television, telecommunications networks and networked computers are seamlessly integrated in a single household appliance. Set-top boxes and the new network computers from manufacturers like Oracle and Apple are supposed to bring computing power and Internet access to television, while TV tuner cards incorporate television with computers.

There has been some success with both, and in the spring of 1997 Microsoft committed itself to developing technology that will bring television to computer desktops while at the same time acquiring WebTV, a company whose goal is to bring the Internet to television sets. At first glance, it might look like the two technologies are mutually exclusive; after all, with the former users watching television on expensive computers, while WebTV allows TV viewers to surf the Net on a comparatively cheap television. However, Microsoft is hedging its bets, figuring that, if one or the other technology succeeds, it will be a goldmine. After all, over the last few years, entertainment moguls and high-technology empires have been drawing closer and closer together in what can best be termed *corporate* convergence.

Throughout the 1990s companies like Lucasfilms and Time-Warner began to explore alliances with the biggest players in the information technology industries, and computer companies courted broadcasters. In 1996 Microsoft began a joint venture with NBC to create MSNBC—a traditional television network, delivered by cable and satellite, with an associated Web site. At most, this trend is no more than a corporate convergence. The medium hasn't changed; new players from the computer industry have plunged into the media business, and media moguls have developed in interest in computers. By 1996 all the major television networks had established Internet footholds. MSNBC is significant only as the archetype of corporate convergence, and because the world's largest software company has diversified its operations.

In the wake of all this Microsoft completely changed its Internet content strategy. The MSNBC partnership remains, but Microsoft has aggressively developed a service that it hopes will revolutionize the Net. The Microsoft Network, created out of the ashes of an ill-fated venture into proprietary on-line services, is intended to bring the familiarity and friendliness of television to the World Wide Web. "We're *not* just putting TV on the Web," insisted Ken Nickerson, MSN Canada's general manager, shortly before the new service was launched in the fall of 1996. "This is an entirely new medium, and at the heart of it is a dichotomy. You have a keyboard and a mouse saying 'touch me,' and a video saying 'sit back and watch me.' We're using the metaphor and creative content of television to do something new. It will have TV's impact and creative content, but unlike television, the user will interact directly *with* that content."

The new MSN and technologies like WebTV and Intel's Intercast don't represent convergence, but something new. McLuhan observed that "what is actually visible in new situations is the ghost of old ones,"[11] and though the so-called convergence technologies allow television to *appear* on networked computers in the same way that movies appear on TV, the *medium* is the Internet. Indeed, new media do not converge with old media; they appropriate part of the old media's functions for themselves, and of all our traditional information media, television has the most to lose in the information revolution. The golden age of radio drama ended abruptly when television—a medium better suited to dramatic content—emerged in the 1940s and 1950s. Just as television cannibalized radio in its early days, the Internet and related interactive media have begun what promises to be a rapid and dramatic erosion of the primacy of television.

> A new medium is never an addition to an old one, nor does it leave the old one in peace. It never ceases to oppress the older media until it finds new shapes and positions for them.[12]

While television is ideally suited to the distribution of dramatic entertainment content, the Internet is not, at least for the foreseeable future. Full-screen video requires the whole bandwidth of the current North American telecommunications backbone,[13] and the implementation of broad-bandwidth consumer technologies has been so slow that it is unlikely that the Internet will be a viable medium for the delivery of video-intensive content like entertainment programming for some time to come. Moreover, the cultural role of the computer as appliance makes it unsuitable as a conduit for informal, even social entertainment. While televisions have occupied pride of place in North American living rooms for decades, computers are typically found in offices and work areas. Whatever entertainment role the computer may have—as a games station or interactive storyteller—it remains primarily an office machine, and a distribution medium for information.

Television will likely retain its position of pre-eminence as a distribution medium for entertainment content, while being stripped of its role as a distribution mechanism for information and news. In 1996 it became clear that Internet users were abandoning television as an information medium. In its *Electronic Access '96* study, Coopers & Lybrand found that 58 percent of Internet users surveyed had shifted their television viewing time directly to the Internet and on-line services. The new media of the Internet and on-line information services undoubtedly provide something that television either doesn't or cannot offer.

Television has never been more than barely adequate as an information medium. The realities of corporate control, sponsorship and scheduling have generally meant that news coverage can never transcend the superficialities of the fifteen-second soundbyte. It doesn't traffic in narratives, but in headlines, with 'pictures at eleven.' In information terms, it is an extremely low-definition medium, forcing viewers to interpolate their own experiences to fill the gaps in the information. More significantly, the dominant model of televisual interaction, in which a consumer reacts to pre-processed information, implies that the interaction is emotional rather than intellectual; it is reactive and cool rather than cognitive and hot.

That was certainly on the minds of NBC's top news executives throughout 1996. By the beginning of 1997, the network's six o'clock newscast finally displaced ABC News as the top-rated network news program. Ironically, this was achieved by actually *decreasing* the amount of news carried on the program. The *new* news featured much less international and political coverage, and longer, human-interest oriented set-pieces. One night in the winter of 1997, when ABC and CBS led with reports of the Whitewater scandal then gripping

Washington, NBC ran a visually-arresting piece on the repair of the Hubble Telescope, and a report on what might happen if an asteroid struck the Earth. The latter story was a tie-in intended to promote a dramatic—and fictional—miniseries on the same network in which just such a catastrophe happened. It all made for great television, but it also underscored the unsettling truth that news and information is *not* great television. As Derrick de Kerckhove has it, "television talks primarily to the body, not to the mind."[14]

The Internet—and interactive digital media in general—talks to both, and thus represents a profound tension between McLuhan's formulation of hot and cool media.[15] On one hand, it is a text-based, high-definition, data-rich medium in which the participant consumes without requiring any interpolation of additional information. Indeed, it is exactly this hot-ness that is being developed and promoted by corporate content providers and on-line publishers. On the other hand, it's cool because, as in television—a data-poor medium—the participant is *involved* in the information process; as much a producer as a consumer of information, filling in the blanks as if watching television, or actively seeking the background of, or context for, a newsbyte. Unlike newspapers, response is not limited or constrained by the goodwill of the letters page editor. Just what appearance the medium acquires is relative to the context of the interaction. Reading a news report is hot; posting a critical Usenet message or providing an alternative Web site substantially cools the medium. Moreover, it is exactly that context or potential that defines the user experience, and consequently, rather than cannibalizing hot media like newspapers and books, the Internet's principal tendency is to subvert cool media like television.

It is just this adaptability that makes the Internet the ideal medium for news. With its unique combination of hot text and cool interaction (not to mention community and social space) the Internet delivers an information consumption experience that is at the same time more interactive and of a higher information resolution than traditional television. The medium conforms to the user's information needs, providing a three-dimensional stage for the news, rather than a flat video proscenium. At the very foundation of Internet interaction is a strong tendency to decentralization, and it remains to be seen whether television's corporate structure will re-make its cannibalized body and resist the current of the datastream.

CHAPTER NINE

Ecstasy and Dread

I. John Brown's Mail Bomb

IMPASSIVE, but with the intensity of an Old Testament prophet, Theodore Kaczynski was led out of a Montana courthouse in handcuffs, a professorial tweed jacket draped over his bright orange prison uniform. This was the man accused of being the Unabomber, one of the most elusive serial killers in American history, whose one-man anti-technology terror campaign had taken the lives of three people and injured 23 others. With his wild hair and unruly beard, Kaczynski looked every bit a latter-day apparition of John Brown, a fanatical terrorist of an earlier age, whose bloody attack on Harper's Ferry in 1860 ignited the American Civil War. Only Kaczynski—if he was the Unabomber—had not pursued his terror campaign to rescue slaves from horrific servitude, but to deliver all humanity from industry, technology and the information revolution.

At the time of Kaczynski's arrest, on April 3, 1996 at his cabin near Stemple Pass in rural Montana, the Unabomber had been a nameless, faceless angel of death who had used the postal system for eighteen years to deliver lethal explosives to airline executives and university professors. The arrest gave the threat a face—a misfit former university professor, an oddball intellectual who dropped out of the 20th century to live in the Montana hills.

The Unabomber's manifesto, published in full by *The New York Times* and *The Washington Post* in September 1995, had already put the mayhem in context and explained its motivation. The Unabomber's campaign was a crusade against the modern world, against technology and industry. The manifesto made it clear that he believed he was pursuing a war of human liberation against the machine.

> The continued development of technology will worsen the situation. It will certainly subject human beings to greater indignities and inflict greater damage on the natural world, it will probably lead to greater social disruption and psychological suffering, and it may lead to

increased physical suffering even in "advanced" countries.[1]

The document was rambling and disjointed, but managed to establish several salient themes. The author was an anarchist, opposed to government and social organization. Large-scale, complex technology, he argued, like industrial machines and computers, dehumanizes individuals by robbing them their autonomy. In order to develop and use technologies any more complex than stone knives and bearskins, individuals must surrender their decision-making authority to the group. After all, he observed, "production depends on the cooperation of very large numbers of people."[2] It is just that cooperation—the imperative of society—that he was rebelling against. The Unabomber's anarchism was more radical and far-reaching than the visions of Bakunin and Kropotkin. While the latter reconciled society and industry with non-hierarchical social structures, the Unabomber's emphasis was on a naive conception of the noble savage.

For the Unabomber, the technological society's greatest crime is the disconnection of humanity from its natural roots. He is a crypto-eugenicist who believes that the modern medical techniques that prolong lives are not only part of the technological conspiracy to destroy individual autonomy, but have corrupted human biology and interfered with the process of natural selection. Above all, he is a primitivist-moralist, elevating the noble savage to a level of holy virtue. While he argued that the anti-technological revolutionary must employ technology—like modern explosives—to destroy the modern world, all technologies that alter the human environment, or require the edifice of modern civilization are evil. Human happiness, he argued, is not found in low infant-mortality rates, extended life spans or education, but in the sylvan utopia of an idealized non-technological state of grace. Nature is a the stable framework, the womb in which humanity, as a collection of atomized individuals, is nurtured in security. In contrast, modern society, in which human society dominates nature, is inherently unstable, due to the pressures of growth and technological change. The change has to be stopped and rolled back, but the Unabomber had no program, no plan beyond killing and maiming those he believed to be the agents of technology and modernity. He was—in his mind at least—going to lead an exodus from the twentieth century.

The aimless ravings of a homicidal maniac, the Unabomber's manifesto struck a chord of resonance in a small but vocal group of anti-technology activists who style themselves neo-Luddites. Taking inspiration from the 19th-century workers who fought a desperate war against technology and the industrial revolution, the latter-day Luddites stand staunchly in the path of technological innovation. They represent a broad spectrum of beliefs—some are willing to accept certain levels of industrial technology as a necessary evil, while others espouse Unabomber-style primitivist-moralism. Not unexpectedly, most neo-Luddites went to great lengths to distance themselves from the Unabomber's terrorism. They may share the same goals, they may be motivated by the same fears, and the terrorist may have put their cause on the front pages of the world's newspapers, but murder and mayhem seem beyond the pale. Yet author Kirkpatrick Sale, the neo-Luddites' most prominent and eloquent trend-setter, also emerged as the Unabomber's principal apologist.

Shortly after the Unabomber's Manifesto became public in September, 1995, Sale wrote "The Unabomber's Secret Treatise: Is there Method in his Madness?" Though he went to great lengths to distance himself from the terrorist's penchant for maiming and killing, Sale answered his question with an unequivocal yes. Indeed, he gave the Unabomber a pat on the back, a tacit vote of support, for bringing the Luddite cause to the front pages of America's newspapers and the lead position on network newscasts.[3] For Sale, the Unabomber was a comrade-in-arms, or at least a fellow-traveller. He saw the terrorist, though rightly reviled for his violence, as a misunderstood prophet of technological doom, a legitimate revolutionary whose diatribe was "persuasive" and "buttressed with careful arguments."

Despite protests that he did not condone the Unabomber's violent methods Sale saw reason in his anti-everything message. To be fair, the writer expressed grave concerns about the terrorist's tunnel vision and amorality. Yet he couldn't repress the admiration he seemed to feel for the Unabomber's magnum opus.

> It is the statement of a rational and serious man, deeply committed to his cause, who has given a great deal of thought to his work and a great deal of time to the expression of it. He is prescient and clear about the nature of the society we live in, what its purposes and methods are, and how it uses its array of technologies to serve them; he understands the misery and anxiety and constriction it creates for the individual and the wider dangers it poses for society and the earth.[4]

By calling the manifesto rational and serious, and commending the Unabomber's prescience, Sale affixed his tacit seal of approval. He seemed to say, implicitly, that this was the kind of diatribe that he would have written himself, and accepted it into the neo-Luddite canon. Indeed, Sale's most biting criticism is that the terrorist failed to ground himself in the Luddistic tradition. That is, the Unabomber was outside of the neo-Luddite orthodoxy that Sale represents.

II. Ned Ludd's Hammer

It is inevitable that periods of rapid and profound technological change are met with resistance and, occasionally, violence. The introduction of the printing press to Europe—and with it, the first widely available editions of the Bible in vernacular translations—was accompanied by a widespread reevaluation of religious orthodoxy, the reformation and religious wars that wracked the continent for generations. The European Industrial Revolution and the introduction of large-scale manufacture based on the power of steam created immense social and demographic dislocations as workers crowded into the industrial centres of Lancashire, London, Paris and the Ruhr. Throughout the early part of the nineteenth-century, the traditional, small-scale craft industries, in which a single craftsman was responsible for the entire process of production, whether it was cloth woven on a loom or iron cast in workshop sands, quickly gave way to the factories' economies of scale. The labour of the highly-skilled craftsman was soon devalued by the innovations of mass production that brought armies of unskilled men, women and children to guide the mechanized production lines.

Their resistance to the new economic and technological order took the form of sporadic attacks on the means of industrial production and its owners. For thirty years, from about 1810 to 1840, the obsolescent labour aristocracy, whose economic power and social privilege had been subverted by steam looms and high-production mills, rallied around General Ludd, the mythical leader of the war on machines and the deliverer of cottage industry. They took hammers to the source of their fear—though, ironically, the Luddites of the 1830s scrupulously avoided doing harm to Charles Babbage or his works—and attacked those mill owners whom they believed were robbing them of their livelihood.

With some notable exceptions—like the Unabomber—today's Luddites are an altogether more peaceable lot. Sale frequently repeats a performance in which he wrecks a personal computer with a sledge hammer, but, for the most part, the neo-Luddites are content to write articles and publish books publicizing their complaints. The focus of their outrage is not so much the mechanized processes of industrial production—though some of the more radical, like Sale, are strong believers in a mythic natural state of grace that humanity has lost through its technological works—but the information revolution. The computer, they argue, "cheapens thought," and dehumanizes the user. In his otherwise thought-provoking book, *The Cult of Information*, Theodore Roszak falls into the neo-Luddites' typically reductionist way of thought. While hearkening to a romanticized notion of education in the past—and, incidentally, ignoring the fact that it was until very recently a privilege of the tiniest minority of the population—Roszak argues that our technological society has reduced learning to data processing, shifting its focus from the world of ideas. Such emphasis, he maintains, fails to recognize the essential godlike characteristic of the human mind, "its inexhaustible potentiality."

In as much as the exaltation of raw data over the inexhaustible potententiality of the mind is an intellectual and social dead end, Roszak is right—but that is almost completely irrelevant. The definitions, based on the binary opposites of *information* and *ideas,* are the neo-Luddites' own construction. They might as well oppose ideas with written language, or faith with the Bible. Indeed, the existence of information necessarily exclusive with ideas and thought, exists nowhere *except* in the neo-Luddites' tautology. In effect, they are incapable of seeing the forest for the trees.

Nowhere is this more true than in the ramblings of neo-Luddite media darling Clifford Stoll. He is the author of *The Cuckoo's Egg*, a gripping account of tracking down a gang of German hackers, in 1989. Stoll is something special to the neo-Luddite cause. He's a technological apostate. Often described as an Internet pioneer—though this is only true in the sense that a switchman in 19th-century Indiana was a railroad pioneer—Stoll led the charge in 1995 against the information revolution.

> Perhaps our networked world isn't a universal doorway to freedom. Might it be a distraction from reality? An ostrich hole to divert our

> attention and resources from social problems? A misuse of technology that encourages passive rather than active participation? I'm starting to ask questions like this, and I'm not the first.[6]

Stoll's principal preoccupation in *Silicon Snake Oil* is the hype that accompanies the information revolution. Perhaps he's right, and there simply *has* been too much hype. However, his great error—and the one that renders his arguments meaningless—is that he mistakes the hype for the reality. Rather than examine how the technology is *used*, he—like the Madison Avenue marketers whom he criticizes—focused almost exclusively on how it's *sold*.

Silicon Snake Oil, is a fine example of the neo-Luddites' universal tendency to fetishize the machine. Reflecting the attitudes of earlier generations who blamed billing errors and technological snafus on The Computer rather than on its operators, they see information technology as a discrete organism, a self-propelled digital hydra, rather than as a tool or a medium. Typically, Stoll mistakes his own romantic preference for handwritten correspondence for an absolute truth about the value of electronic mail.

> Network mail, even decade old e-mail, lacks warmth. The paper doesn't age, the signatures don't fade... Real mail has pretty stamps and postmarks—foreign envelopes show mysterious pictographs that tempt me to visit. There's a return address, reminding me of a friend's home that I once visited. Here's a pressed flower from some faraway summer. These letters bring back promises, memories, and smiles.[7]

Stoll conflates the personal with the universal, the subjective with the objective. His book begins with an anecdote about being addicted to the Internet; how he is too wrapped up in his electronic correspondence to join his friends—who he evidently invited to his home and then promptly ignored—watching a basketball game. From there, he magically transforms his own poor manners into a judgment on the value of the information revolution. The rationale appears to be "Cliff Stoll needs a life, and is too lazy to get up from the computer, so the whole information revolution must be a sham."

Stoll is careful to mention in radio and television appearances, and throughout his book, that he really *loves* computers, and that he has worked in university computer departments so, damn it, *he should know*. Indeed, that is his principal value to the neo-Luddite cause. His position as a technological apostate seems to lend him an air of authority, even though all of his criticisms are based on his personal shortcomings. The recurrent neo-Luddite refrain that information technology somehow undermines literacy seems all the more persuasive coming from a former member of the academic computing community. For example, Stoll warns that computers and information technologies are cannibalizing libraries, and that the resulting institutions won't simply be libraries without books, but "a library without value."[8] Yet, just as with e-mail, this is a statement of fetish and not reality. Stoll confuses the *medium* of communication with *communication*, he can't tell the difference between the *form* of a book with the *content* of a book. What difference does it make whether a work of literature is carved in stone, etched on a scroll, printed in a leather-bound volume or displayed on the phosphors of a cathode-ray tube? Information on the Internet, whether it's communiqués from the Mexican jungles, newspapers or literature, is encoded as text. Indeed, the information revolution has made literacy an even more vital skill that it ever was before. The principal difference is that access to this information is more open than ever... and that is exactly the problem.

As a group, the leading neo-Luddites are remarkably homogeneous. They tend to be men, and survivors of the 1960s counterculture—former hippies whose attitudes toward technology were formed at a time when computers and data communications networks were tools of the military-industrial complex. It isn't much of a step from protesting the technological Big Brother outside campus computer buildings to blaming the information revolution for everything that's wrong with the world. Even more significant is the fact that almost all of them are comfortable, middle-class intellectuals, either established writers or tenured university professors. The neo-Luddites' identification with the dispossessed weavers of 19th-century Lancashire, is spurious at the very least. The historical Luddites were workers, skilled craftsmen whose livelihoods were directly threatened by the technological innovations of the Industrial Revolution. Steam looms and industrial weaving put them out of work, and their fear of technology, though misguided, had its foundation in reality.

In contrast the information revolution is unlikely to force Sale from his Greenwich Village brownstone or Stoll from his Oakland, California, home and into a life of squalor. The only thing they have to lose is social position and prestige. These are the traditional guardians of knowledge and information of a society where access to information is restricted by class and circumstance. If the tools of information are available for everyone to use, then people like Sale, Roszak and Stoll—as paternalistic custodians of information—are socially irrelevant. This is the basis of Roszak's contrast of ideas and information. The foundation of this formulation is the assumption that *someone*, some intellectual authority must direct the synthesis of information (or images) in ideas. "Master ideas are cheapened when they are placed in the keeping of small, timid minds that have grown away from their own childish exuberance."[9] Conversely, in the hands of great artists "like Homer," they retain their complexity and greatness. Great ideas, Roszak says, are for great men and women. In artistic terms, this may be true, but the judgment of *who* has the ability to manage and manipulate images is ideological rather than objective. The information revolution potentially puts images and discretely organized parcels of data in the hands of everyone; and many, if not most, of us fall far short of the intellectual powers of a Homer, or even a Theodore Roszak. Whether that 'cheapens thought' or ill serves society, depends on how much authority over information and ideas one stands to lose.

III. Wishful Thinking

There was once a future in which the marvels of modern science and engineering would answer all of our needs. *Popular Mechanics* and *Popular Science* promised us hovercars, video-phones, and self-navigating automobiles. Our cities were to be high-tech paradises—a pastiche of Fritz Lang, Corbusier, Frank Lloyd Wright and Leo Gernsback's pulp science fiction. We would lead better lives through science, engineering and George Jetson gadgets.

And I'm still waiting.

We are always obsessed with the future during periods of rapid technological change. Manifestations of human ingenuity inevitably seem to presage some distant social and economic bounty that remains just beyond our grasp beyond the uncharted new horizons promised by the exhibits at World's Fairs in the 1940s and 1950s. The current technological upheaval is no different, and secular prophets, like their biblical prototypes in collapsing Israel, have sprung up everywhere. They are futurists, technophiles who see the clear paths to the future etched in the tools of the information revolution. In the chaos of technological change and its attendant impact on our society and the economy, they see the trends of revolution playing out with the inevitability of an elaborately choreographed ballet. Futurists (or futurologists) are not, of course, a creation of the information revolution, but it is in this current period of rapid transformation that their vocation has met apotheosis.

> ... futurology is now more pervasive than ever. And modern prophecy has entered a new phase, which sometimes makes its effects more subtle. Like all charismatic movements, it has finally become institutionalized and rationalized, has become a given, a presence, an essential part of our corporate and bureaucratic life and culture, as well as of our personal lives.[10]

The irony is that, if the information revolution has demonstrated anything, it is that prophecy, even long-term forecasts, are utterly worthless. The evolution of information technologies has been a process of contingency following on contingency. Tools developed for one purpose have been subverted by others. Who, for example, could have imagined that ARPANET, developed as a Cold War research project, would become a global mass medium, the scale of which we have never seen before?

Yet the future and futurism are integral parts of the corporate ideology, and to a large extent it is the corporations that attempt to set the agenda of the information revolution. In his 1995 book *The Road Ahead*, Bill Gates, the Microsoft CEO and arguably the single most influential corporate figure in the information economy, gazed into the future and confidently predicted that the new world will be a paradise. While Gates extrapolated the observable trend in the information revolution to home-based work and the decentralization of corporate activities, his conclusion that this will somehow enrich our lives does not *necessarily* follow. As we have seen, there are all kinds of new pressures, stresses, and even physical dangers implicit in telework.

Taken in isolation, the new technologies of work may seem to hearken to the same kind of golden age promised by General Motors in 1939, but socio-economic changes are far more complex than that. Writing of James Burnham's political forecasts in the 1940s, George Orwell observed that "political predictions are usually wrong, because they are usually based on wish-thinking."[11] The same can easily be said of technological predictions, but even more revealing, in this case, is *who* made the predictions, and what their agenda was.

The proliferation of prophets isn't surprising given the pace of the technological change attending the information revolution. The transformations have been dizzying, and they appear to come more quickly than we can assimilate them. Just when you get used to your new workplace, some new technology comes along and changes everything. Technological obsolescence has become a fact of life, and the *fear* of obsolescence has become a driving force of the new economy. Strategic planning, the imperative to "stay ahead of the curve," has created a fertile environment for those prophets who claim to have all the answers. In 1996, Andy Grove, the CEO of Intel, one of the most powerful corporations of the information economy, wrote:

> It's like sailing a boat when the wind shifts on you but for some reason, maybe because you are down below, you don't even sense that the wind has changed until the boat suddenly heels over. What worked before doesn't work anymore; you need to steer the boat in a different direction quickly before you are in trouble, yet you have to get a feel of the new direction and the strength of the wind before you can hope to right the boat and set a new course.[12]

Faced with the uncharted, and potentially treacherous waters, we have created a near limitless demand for navigators with preternatural knowledge of the unknown and unknowable shoals and reefs of the future. One corporate axiom has it that 'staying ahead means looking ahead,' and that requires the services of a prophet.

In most respects, futurism is little different from science fiction—science fiction writers simply use the future to satirize and critique the present, while the futurist extrapolates the present to see the future. His obsession is the trend, or more specifically, the megatrend, a term coined by futurist John Naisbitt in 1982 to describe global, decade-long transformations that can be forecast and conveniently categorized.[13] Yet the truth is that these megatrends are arbitrary, simplistic and superficial by definition. Naisbitt and his colleagues seek to extrapolate a vision of the future from current conditions without having any way of accounting for the near infinite possibilities and permutations of possibilities in the real world. Whether his forecasts are accurate is immaterial—though it is possible to create an appearance of accuracy by keeping things vague—the point is that society and technology are far too complex to be predictable. We can't even predict the weather with any certainty beyond three days, despite the volume of technology thrown at the problem! Indeed, anyone who has ever experimented with a wave tank in a high school physics class knows that it is impossible to model a complex system by the behaviour of a small, closed system.

The power of the futurist's visions is not in what it reveals of the shape of things to come, but how it can impose itself on the future. Megatrends are inevitable because, by establishing them as the agenda and program of the economy, they *become* inevitable. The prophecies are self-fulfilling. If enough people in the computer industry are led to believe that the wave of the future will be ultra-compact computers, then that is exactly what will happen. In their rush to capitalize on the coming megatrend, the sheer momentum will make the prophecy come to pass. In the entertainment and retail industries, they call it hype or buzz. During the 1996 Christmas shopping season, so many television celebrities appeared on camera with the Tickle-me Elmo dolls, commenting on how it was the hottest toy of the season that, in fact, it *became* the hottest toy of the season.

Futurism is nothing less than ideological marketing. It claims to show the future of technology and society in order to set course for that specific future. Futurists and strategic planners—the darlings of the corporate economy—are simply corporate ideologues, setting the goals and agenda of business for the next ten, 20 or 50 years. Prophecy inevitably serves ideology. By promising bright futures, ideologues give their five-year plans and social projects a material goal, and by presenting their concrete programs as objective results of an inevitable process, they give their goals the stamp of natural law. It is part wishful thinking and part concrete socio-economic plan, reflecting the interests, goals and preoccupations of their corporate sponsors.

Naisbitt's prophecies for the last decade of the millennium are particularly revealing in this regard. His forecasts drip with enthusiasm for the doctrine of free trade and the global extension of the neo-liberal economic order, and read more convincingly as prescriptions than as predictions. Ideological and trade barriers to world commerce *will* erode, in his thinking, because they *should* erode. Built into these prescriptions is the neo-liberal principle that corporate economic growth will benefit the whole world. Echoing the futurist sentiments of General Motors' 'Futurama' exhibit at the 1939 World's Fair, Naisbitt declares that "the developed world's economic boom will be the foundation for higher evolution and global affluence."[14] And this economic boom (seen in appropriately messianic terms by a secular prophet) is built on a 'triumph of the individual' fostered by the information revolution. However, his triumphant individuals—often independent professionals displaced by a decade of corporate downsizing and outsourcing—exist only as consumers in his formulation of the information revolution. This is the essential thrust of corporate futurism. The subtext is not so much that the *customer* is king, but that the product, whether information or information technology tools, reigns supreme in the information revolution. Implicit in Naisbitt's prescriptive predictions, and in the futurist position generally, is the traditional belief in 'better living through science.' The path to human social and spiritual liberation lies through the consumption and acquisition of machines. Utopia will be technological.

Techno-utopianism, a dominant subset of futurism which holds that the enormous and accelerated technological developments of the information revolution will dramatically improve human life and society is less about utopia than about technology. It is as much a fetish as the neo-Luddites' repudiation of all of the works of humanity, though guided by a different ideology. It's a new spin on Fordism, but on a social scale, in almost all respects little different from the confident techno-industrial ideology expounded by Ford and General Motors in the 1930s and 1940s, and it underlies the neo-liberal ideology of the corporate information economy. Its principal theme, is that consumption will make you free, that the products of an advanced economy—in this case, information products—are the key to future advancements in human comforts and quality of life. In his book, Gates' twin preoccupations are the high-tech Xanadu he is building outside of Seattle, and the products of the future—whether they are information services, software or wallet computers—that Microsoft plans to develop and sell. The subtext of his thesis is that what's good for business, and thus what is good for Microsoft, is good for society as a whole.

Significantly, this view has become the ideology of the corporate neo-liberal elite and the governments it sponsors. In social terms, we are to be given products rather than tools, and rather than being participants in the global economy, we will be consumers. At the level of government, statecraft becomes management—nothing underscores the influence of this corporate hegemony as well as the obsession that governments of every industrialized country have with deficit cutting and fiscal management. The desire to put one's house in order is noble, but the proponents of neo-liberal managerialism haven't quite explained what the ultimate goal of the cuts will be. Deficit reduction is its own goal, implicitly for the maintenance of a healthier commercial climate for the concentration of capital and control of commodities, and it's hard to escape the conclusion that the state is becoming the local branch manager for the transnationals.

IV. The Fetish Machine

It's an inescapable irony of the information revolution that both the neo-Luddites and the techno-utopians have a lot more in common that either would be willing to admit. In 1995, I was invited to participate on a national CBC Radio program about the Internet, and found myself sandwiched between opposing sides of the debate. John Perry Barlow, the former Grateful Dead member who co-founded the Electronic Frontier Foundation, spoke live from distant Australia, waxing poetic about the Net and eagerly anticipating the time in the presumably not-too-distant future when we would be able to connect to cyberspace through some kind of neuro-electronic interface. He spoke of a peculiar deus ex machina, of how the machine was poised to effect the apotheosis of humanity. Granted a pre-recorded right of rebuttal, Clifford Stoll, not unexpectedly, took the opposite position. The technology is evil, he contended, it robs us of our humanity, and, before we know it, we will be interacting solely with machines, and not with other human beings. For Barlow, the technology is a species of god; for Stoll, it is a demon.

One need only strip away the moralistic value judgments to see that they were saying essentially the same thing. The techno-utopian's god is the neo-Luddite's demon, but in either case, technology is presented as an object alien to human works, needs and society. It is a fetish object. In neither equation is the human being or human *use* of the technology considered. Techno-utopianism and Luddism are profoundly dehumanizing; the former because it reduces human society to little more than a convenient concentration of consumers and the latter in its inability to see beyond a sentimental and idealized conception of a noble savage corrupted by his tools.

V. Technology R Us

The Cyberpunk science fiction stories of William Gibson and Bruce Sterling take place in a world of pirates and thugs, of vastly powerful transnational corporations and fashion conscious kids. In other words, it's a world not too different from our own. The stories are satires of the information revolution and the emerging global economy. "It's only in North America that I've been taken absolutely seriously... Japan as well," Gibson said in 1995. "The British and the French assumed from the beginning that I was to some very large extent a humorist, and that's from *Neuromancer* on. There are things in *Neuromancer* that were very deliberately written to be quite funny." Nevertheless, at least part of the satirist's métier is deadly serious. It's a reality check, a report on the state of society and the gravity of its weaknesses.

Technology, or rather its place and role in human society is Gibson's central preoccupation. His heroes struggle to maintain their humanity in the face of a constantly-shifting definition of human. Technology is both a threat and a promise, as the line between man and machine blurs to irrelevance. The dominant theme is that humanity is a measure of how we integrate our works and tools into our society and ourselves. Gibson sees no distinction between the human and the technological. Humanity is a process of synthesis, from the earliest stone tools to the Net. "It always just seemed to be to me so much of what we're about as human beings," he said. "Certainly in the time that I've been alive. It's not like there's 'us,' and then there's the technology. Technology 'R Us. Already we're sort of half machine." His characters are technological creatures, "but not really much more than we are. We're supposed to have children at age fourteen and pretty much be dead by thirty. We don't do that. Technology's not something you can put back in the black box and return to Radio Shack. It's the vaccinations you've had and all that metal in your teeth. It's everything, the whole thing that makes us what we are. There is absolutely no option. Whatever it is, we're going there. We can't say 'Oh no, back to nature!'

because nature no longer exists. We messed with it too much."

Both technophilia and technophobia—in the form of the futurist and Luddite positions, respectively—place technology beyond the human experience. The machine is either a demon or a god, but either way it denies humanity's place within the context of its own works. The distinction is not a new one; our species is a tool maker, and to deny or deify those tools degrades our status and counsels dangerous indifference. The binary opposites of man and machine, of natural and technological are thus an abdication of humanity. We are, indeed, half-machine, in the process of integrating our technologies into our lives and experience—and this is the essential process of the information revolution.

The information revolution has fundamentally altered the way we work and interact, broadened the theatre of international politics and begun to redefine the position of the individual in civil society. It has spawned a new global economy founded on the exchange of intangibles—and it has made a few people very, very rich, yet it is so far, a passive revolution, which has left the fundamental socio-economic relations untouched. We may work differently, but the *context* of our work—that is, our position within the means of production—is unaltered despite the appearance of a new product.

Beyond that, it is a revolution of possibilities in which the final outcome is still up for grabs. We can no more predict the result than an 18th century Lancanshire weaver could have known how the Industrial Revolution would change his world. It's too early to tell where any of the information revolution's many threads will ultimately lead. The changes will be profound and numerous. They will require new eyes and new language to describe and comprehend, and they will doubtless improve our lives at the same time as they make them more difficult.

So far, it has been a revolution with three protagonists—the state, the corporate economy and the users—exerting contrary forces in opposing directions, holding out both promises and perils. Just as there is the promise of information liberation implicit in the evolution of the unique and parallel social space of the Internet and in the rapid social and geographic distribution of the means of information production, there is also the danger that we are rushing headlong into a trans-national monopoly capitalism that respects no borders and answers to no vote. The information and knowledge of the world has become, though the technological extensions of the human mind, a global common resource, yet the processes that have made this possible require that information remain a scarce and controlled commodity. And in desperate response to the challenge to find relevance in a world that knows no borders for capital or information, the state twists and writhes, arbitrarily enforcing its authority here while abdicating to the forces of capital there. Government policies, whether they are latter-day blue laws or xenophobic cultural proscriptions, only contribute to the scarcity of information, and thus its commodification and concentration by transnational interests. Yet free access to information will, by definition, subvert the authority of the state and, more importantly, undermine the economic *necessity* of the information revolution.

How we respond to the challenges of the information revolution—as individuals, as countries and as a global community united in a single economy—will ultimately determine its course. Its tools have the potential to be truly liberating technologies, but the potential has a limited shelf-life. We can neither afford to recoil in horror and fear at the changes that have only just begun, nor be overwhelmed by the intoxicating promises of a consumer utopia. Faced by the objective necessity of history, the emergent global society must act for itself as the sole protagonist of the information revolution. The street must make its own use of things.

Notes

NOTES TO INTRODUCTION

[1]Gibson, William, *Neuromancer* (New York: Ace Books, 1984, 1994) 51.
[2]Orwell, George, *A Collection of Essays* (San Diego: 1946, 1981) 311.

NOTES TO CHAPTER ONE

[1]Cortada, James, *Before the Computer* (Princeton: Princeton University Press, 1993) 7.
[2]Lovelace, Ed. Note to Menabrea, L.F., "Sketch of the Analytical Engine Invented by *Charles Babbage*, esq" in Bowden, B.V., *Faster than Thought* (New York: Pitman, 1953) 349.
[3]Ibid, 368-369.
[4]Bowden, B.V., *Faster than Thought* (New York: Pitman, 1953) 17.
[5]Hodges, Andrew, *Turing: The Enigma* (London: Burnett, 1983) 109.
[6]Ibid, 109.
[7]Bowden, 131.
[8]Dickson, Paul, *Think Tanks* (New York: Atheneum, 1972) 9-10.
[9]Russell, Cheryl, "High-Tech Hoopla," *American Demographics* (June 1983) 34.
[10]Tapscott, Don, *The Digital Economy: Promise and Peril in the Age of Networked Intelligence* (New York: McGraw-Hill, 1996) 6.
[11]Mowlana, Hamid, *Global Information and World Communication: New Frontiers in International Relations* (New York: Longman, 1986) 94.
[12]Tapscott, 7.
[13]Levy, Steven, *Hackers* (New York: Dell, 1984).
[14]Schiller, Herbert I., *Information Inequality: the Deepening Social Crisis in America* (New York: Routledge, 1996) 35.

NOTES TO CHAPTER TWO

[1]Hafner, Katie, and Lyon, Matthew, *Where Wizards Stay up Late: The Origins of the Internet* (New York: Simon & Schuster, 1996) 43-44.
[2]Baran actually called the system "distributed adaptive message block switching." The Welsh researcher Donald Davies, who was developing essentially the same concept at the same time, coined the phrase "packet switching."
[3]Hafner and Lyon, 190-191.
[4]Rheingold, Howard, *The Virtual Community: Homesteading on the Electronic Frontier*, (New York: HarperCollins, 1994, http://www.minds.com/) 24.
[5]Logan, Robert K., *The Fifth Language: Learning a Living in the Computer Age* (Toronto: Stoddart, 1995) 273.
[6]Rheingold, 25.

[7]Cherney, Lynn and Weise, Elizabeth Reba, eds., *Wired Women* (Seattle: Seal Press, 1996) xii.

[8]Oldenburg, Ray, *The Great Good Place: Cafes, Coffee Shops, Community Centers, Beauty Parlors, General Stores, Bars, Hangouts, and How they Get You Through the Day* (New York: Paragon House, 1989) 4.

[9]McLuhan, Marshall, *Understanding Media:The Extensions of Man* (New York: McGraw-Hill, 1964) 7.

NOTES TO CHAPTER THREE

[1]Tapscott, Don, *The Digital Economy: Promise and Peril in the Age of Networked Intelligence* (New York: McGraw-Hill, 1996) 6.

[2]Schiller, Herbert, *Information Inequality* (New York: Routledge, 1996) 116.

[3]Connecting the North Symposium Final Report, *Northern Voices on the Information Highway* (Ottawa, 1995) i-ii.

[4]Comité Consultatif sur l'Autoroute de l'Information (Québec), "Inforoute Québec: Plan d'Action Pour la Mise en Oeuvre de l'AutoRoute de l'Information" (Quebec City, 1995) 30.

[5]Comité Consultatif, page 30.

[6]Commission de la Culture (Quebec), "Les Enjeux du Développement de l'Inforoute Québecoise" (Quebec City, 1996).

NOTES TO CHAPTER FOUR

[1]Breslow, Harris, "Locating the Politics of the Internet," unpublished paper, 1996, 6.

[2]Rheingold, 48.

[3]Ross, Andrew, "Hacking Away at the Counterculture," *Technoculture* (Minneapolis: University of Minnesota Press, 1991) 120.

[4]Yates, Larry Lamar, "The Internet: What it Can and Can't Do for Activists," presentation to Inet 96, Montreal, June, 1996.

[5]General Command of the EZLN, *Declaration of War*, San Cristobal, 31 December 1993.

[6]General Command of the EZLN, *Declaration of War.*

[7]General Command of the EZLN, "Second Declaration from the Lacandona Jungle," 10 June 1994, *!Zapatistas! Documents of the New Mexican Revolution*, Autonomedia (Brooklyn, 1994) 330.

[8]Cleaver, Henry, "The Zapatistas and the Electronic Fabric of Struggle," University of Texas (Austin, 1996) http://www.eco.utexas.edu/faculty/Cleaver/Zaps.html

[9]Swett, Charles, "Strategic Assessment: The Internet," Office of the Assistant Secretary of Defense for Special Operations and Low-Intensity Conflict (Washington, 1995) http://www.fas.org/cp/swett.html

[10]Salvaggio, Jerry L., ed., *The Information Society: Economic, Social and Structural Issues* (Hillsdale: Lawrence Erlbaum Associates, 1989) 108.

[11]Department of Defense (U.S.), "The Militarily Critical Technologies list, Part 1: Weapons Systems Technologies" (Washington: Office of the Under Secretary of Defense for Acquisition & Technology, June 1996) 9-1.
[12]Swett.
[13]Stein, George J., "Information War–Cyberwar–Netwar," New Dimensions International, 1996. http://www.cdsar.af.mil/battle/chp6.html.

NOTES TO CHAPTER FIVE

[1]Beam, Louis, "Leaderless Resistance," *The Seditionist*, no. 12 (February 1992).
[2]Keeler, Bob, "Assault on History; Is America fertile ground for those who claim the Holocaust didn't happen? Jewish leaders are hard-pressed with such an absurd proposition," *Newsday*, 24 February 1994.
[3]Smith, Winston (Harold Covington), "The Importance of Holocaust Revisionism," *NSNet Bulletin*, 5 (July 1996).
[4]McVay, Kenneth N., "Holocaust FAQ: Willis Carto and the Institute for Historical Review," 10.
[5]Zundel later consented to an interview, but too late for the present book.
[6]Anti-Defamation League Report, "Hate Group Recruitment on the Internet," 1995.
[7]Kleim, Milton J., "On Tactics and Strategy for Usenet," 1995.
[8]Kleim, Milton J., "National Socialism Primer," Second Edition, 1995.
[9]Nizkor is Hebrew for "we must remember."

NOTES TO CHAPTER SIX

[1]Johnston, David, Johnston, Deborah and Handa, Sunny, *Getting Canada Online: Understanding the Information Highway* (Toronto: Stoddart, 1995) 200.
[2]Rheingold, 257.
[3]Mulgan, G.J., *Communication and Control: Networks and the New Economies of Communication*, Polity (Cambridge: Polity, 1991) 130.
[4]Negroponte, Nicholas, *Being Digital* (New York: Vintage Books, 1995) 58.
[5]This was the claim made in several lawsuits brought by freelancers against newspaper publishers in spring, 1997.
[6]Southam Newspapers, Freelance Agreement, March, 1996.
[7]Simon Wiesenthal Center (Canada), *The Need for Regulation on the Information Highway*, Toronto, 1995) 2.
[8]Ibid, 2.
[9]Ibid, 3.
[10]Littman, Sol, "Some Thoughts on the Regulation of Cyberspace," (1995) 2.
[11]Ibid, 6.
[12]Elmer-Dewitt, Philip, "On a screen near you: Cyberporn," *Time*, 3 July 1995, 38.
[13]Wallace, Jonathan and Mangan, Mark, *Sex Laws and Cyberspace* (New York: Henry Holt 1996), 144.

[14]Distirct Court for the Eastern District of Pennsylvania, ACLU v. Reno, 11 June 1996.

[15D]Ibid

[16]Elmer-Dewitt, 40.

[17]ACLU v. Reno.

[18]Sansom, Gareth, "Illegal and Offensive Content on the Information Highway," (Ottawa, 1995) 12.

NOTES TO CHAPTER SEVEN

[1]Boyett, Joseph H., and Conn, Henry P., *Workplace 2000: The Revolution Reshaping American Business* (New York: Dutton, 1991) 35.

[2]Logan, Robert, *The Fifth Language: Learning a Living in the Computer Age* (Toronto: Stoddart, 1995) 207.

[3]Interview with Susan Di Poce, Lotus Development Canada, October, 1995.

[4]Tenner, Edward, *Why Things Bite Back: Technology and the Revenge of Unintended Consequences* (New York: Alfred A. Knopf, 1996) 173.

[5]Ibid, 175.

[6]Ibid, 175-176.

[7]Boyett and Conn, 45.

[8]Siegel, Lenny, and Markoff, John, *The High Cost of High Tech: The Dark Side of the Chip* (New York: Harper & Row, 1985) 106-107.

[9]Tapscott, Dρn, *The Digital Economy: Promise and Peril in the Age of Networked Intelligence*, New York: McGraw-Hill, 1996) 301.

[10]Logan, 293.

[11]Mandel, Michael J., "The Digital Juggernaut," *BusinessWeek* (Special), 18 May 1994, 28.

[12]Russell, Cheryl, "High-Tech Hoopla," *American Demographics*, (June 1983) 35.

NOTES TO CHAPTER EIGHT

[1]Ontario Court of Justice (General Division), St. Catharines, Ontario, R. v. Bernardo, 5 July 1993.

[2]Resnick, Rosalind, "Banner Backlash," *Internet World,* October, 1996, 10:8.

[3]Negroponte, 153.

[4]Stephens, Mitchell, *A History of the News* (New York: Viking, 1988) 163.

[5]Newspaper Association of America study, http://www.naa.org/.

[6]Ibid.

[7]Logan, 278.

[8]Blanshay, Marlene, "The Paperless Academy," *Quill & Quire*, November 1995, 13.

[9]Coopers & Lybrand L.L.P., *Electronic Access '96*, June, 1996, http://www.colybrand.com/industry/infocom/mediapr.html.

[10]Negroponte, 84-85.

[11]McLuhan, Marshall, "Postures and Impostures of Managers Past," *Essential McLuhan* (Concord, Ont.: Anansi, 1995) 83.

[12]McLuhan, Marshall, *Understanding Media:The Extensions of Man* (NewYork: McGraw-Hill, 1964) 174.
[13]Dejesus, Edmund X., "How the InternetWill Replace Broadcasting," *Byte*, February 1996, 51-52.
[14]de Kerckhove, Derrick, *The Skin of Culture* (Toronto: Somerville House, 1995) 8.
[15]McLuhan, 1964, 22-24.

NOTES TO CHAPTER NINE

[1]FC (The Unabomber), "Industrial Society and its Future," 1995. Paragraph 1.
[2]FC (The Unabomber), paragraph 117.
[3]Sale, Kirkpatrick, "Unabomber's SecretTreatise: IsThere Method in His Madness?" *The Nation*, 25 September 1995, 305.
[4]Ibid, 311.
[5]Roszak,Theodore, *The Cult of Information:A Neo-LudditeTreatise in High-Tech,Artificial Intelligence, and the True Art of Thinking* (Berkeley: University of California Press 1986, 1994) 244.
[6]Stoll, Clifford, *Silicon Snake Oil: Second Thoughts on the Information Highway* (New York: Anchor Books, 1995) 2.
[7]Ibid, 158.
[8]Ibid, 216.
[9]Roszak, 240.
[10]Dublin, Max, *Futurehype: The Tyranny of Prophecy*, Viking (Markham, Ont.: 1989) 53.
[11]Orwell, George, "James Burnham and the Managerial Revolution," *The Collected Essays, Journalism and Letters of George Orwell,Volume IV: In Front of Your Nose,* edited by Sonia Orwell et al., (London, Camelot 1968) 172.
[12]Grove, Andrew S., *Only the Paranoid Survive: How to Exploit the Crisis Points That Challenge Every Company and Career* (New York: Doubleday 1996) 20.
[13]Naisbitt, John, and Aburdene, Patricia, *Megatrends 2000* (New York: William Morrow, 1990) 12.
[14]Naisbitt and Aburdene, 311.

Bibliography

Print Media

Books

Bowden, B.V. (ed.), *Faster Than Thought*, Pitman, New York, 1953 (1964).

Boyett, Joseph H., and Conn, Henry P., *Workplace 2000: The Revolution Reshaping American Business*, Dutton, New York, 1991.

Cherny, Lynn and Weise, Elizabeth Reba (Eds.), *Wired Women: Gender and New Realities in Cyberspace*, Seal Press, Seattle, 1996.

Cortada, James, *Before the Computer*, Princeton University Press, 1993.

Desbarats, Peter, *Guide to Canadian News Media*, Harcourt Brace Jovanovich, Toronto, 1990.

Dickson, Paul, *Think Tanks*, Atheneum, New York, 1971.

Dublin, Max, *Futurehype: The Tyranny of Prophecy*, Viking, Markham, 1989.

Eaman, Ross A., *The Media Society*, Butterworths, Toronto, 1987.

Gates, Bill, *The Road Ahead*, Viking, New York, 1995.

Gibson, William, *Neuromancer*, Ace Books, New York, 1984 (1994).

Grove, Andrew S., *Only the Paranoid Survive*, Currency, Doubleday, New York, 1996.

Hafner, Katie, and Lyon, Matthew, *Where Wizards Stay Up Late: The Origins of the Internet*, Simon & Schuster, New York, 1996.

Hafner, Katie, and Markoff, John, *Cyberpunk: Outlaws and Hackers on the Computer Frontier*, Touchstone, Simon & Schuster, New York, 1991.

Haraway, Donna, *Simians, Cyborgs, and Women: The Reinvention of Nature,* Routledge, Chapman & Hall, New York, 1991.

Hodges, Andrew, *Turing: The Enigma*, Burnett, London, 1983.

Jacobs, Jane, *The Death and Life of Great American Cities*, Vintage, New York, 1971.

Jacoby, Russell, *The Last Intellectuals: American Culture in the Age of Academe*, Basic Books, New York, 1987.

Johnston, David, Johnston, Deborah, and Handa, Sunny, *Getting Canada Online: Understanding the Information Highway*, Stoddart, Toronto, 1995.

de Kerckhove, Derrick, *The Skin of Culture*, Somerville House, Toronto, 1995.

Kinsella, Warren, *Web of Hate: Inside Canada's Far Right Network*, HarperCollins, Toronto, 1994.

Kurtz, Howard, *Media Circus: The Trouble with America's Newspapers*, Times Books, New York, 1993.

Levy, Steven, *Hackers: Heroes of the Computer Revolution,* Dell, New York, 1984.

Logan, Robert K., *The Fifth Language: Learning a Living in the Computer Age*, Stoddart, Toronto, 1995.

Lyotard, Jean-François, *The Postmodern Condition: A Report on Knowledge*, University of Minnesota Press, Minneapolis, 1979 (1984).

McLuhan, Eric, and Zingrone, Frank (Eds.), *Essential McLuhan*, Anansi, Concord, Ont., 1995.

McLuhan, Marshall, *The Gutenberg Galaxy: The Making of Typographic Man*, University of Toronto Press, Toronto, 1962.

McLuhan, Marshall, *Understanding Media: The Extensions of Man*, McGraw-Hill, New York, 1964.

Mowlana, Hamid, *Global Information and World Communication: New Frontiers in International Relations*, Longman, New York, 1986.

Mulgan, G.J., *Communication and Control: Networks and the New Economies of Communication,* Polity, Cambridge, 1991.

Naisbitt, John, and Aburdene, Patricia, *Megatrends 2000*, William Morrow and Company, New York, 1990.

Negroponte, Nicholas, *Being Digital*, Vintage, New York, 1995.

Orwell, George, *A Collection of Essays*, Harvest, San Diego, 1946 (1981).

Orwell, George, *The Collected Essays, Journalism and Letters of George Orwell, Volume IV: In Front of Your Nose*, Camelot, London, 1968.

Penley, Constance, and Ross, Andrew (Eds.), *Technoculture*, University of Minnesota Press, Minneapolis, 1991.

Rheingold, Howard, *The Virtual Community: Homesteading on the Electronic Frontier*, HarperCollins, New York, 1994.

Rose, Frank, *West of Eden: The End of Innocence at Apple Computer*, Viking, New York, 1989.

Roszak, Theodore, *The Cult of Information*, University of California Press, Berkeley, 1986 (1994).

Salvaggio, Jerry L., ed., *The Information Society: Economic, Social and Structural Issues*, Lawrence Erlbaum Associates, Hillsdale, 1989.

Schiller, Herbert I., *Information Inequality*, Routledge, New York, 1996.

Shurkin, Joel, *Engines of the Mind*, W.W. Norton & Co., New York, 1984.

Siegel, Lenny, and Markoff, John, *The High Cost of High Tech: The Dark Side of the Chip*, Harper & Row, 1985.

Sinclair, Carla, *Net Chick: A Smart-Girl Guide to the Wired World*, Henry Holt and Company, New York, 1996.

Stephens, Mitchell, *A History of the News*, Viking, New York, 1988.

Sterling, Bruce, *The Hacker Crackdown: Law and Disorder on the Electronic Frontier*, Bantam, New York, 1992 (1993).

Tapscott, Don, *The Digital Economy: Promise and Peril in the Age of Networked Intelligence*, McGraw-Hill, New York, 1996.

Tenner, Edward, *Why Things Bite Back: Technology and the Revenge of Unintended Consequences*, Knopf, New York, 1996.

Wallace, Jonathan, and Mangan, Mark, *Sex, Lies and Cyberspace: Freedom and Censorship on the Frontiers of the On-Line Revolution*, Henry Holt, New York, 1996.

Weber, C., and Phillip, A., *Computer Professionals in Canada: A Survey of Supply and Demand*, Queens ITC Press, Kingston, 1995.

Papers, Studies and Government Documents

Breslow, Harris, "Locating the Politics of the Internet," York University, Toronto, 1996.

Cleaver, Harry, "The Zapatistas and the Electronic Fabric of Struggle," University

of Texas, Austin, 1996.
Commission de la Culture (Québec), "Les Enjeux du Développement de l'Inforoute Québécoise," Quebec City, 1996.
Comité Consultatif sur l'Autoroute de l'Information (Québec), "Inforoute Québec: Plan d'Action Pour la Mise en Oeuvre de l'Autoroute de l'Information," Quebec City, 1995.
Connecting the North Symposium Final Report (Canada), *Northern Voices on the Information Highway*, Ottawa, 1995.
Coopers & Lybrand L.L.P., *Electronic Access '96*, June, 1996.
Department of Defense (U.S.), "The Militarily Critical Technologies List, Part 1: Weapons Systems Technologies," Office of the Under Secretary of Defense for Acquisition & Technology, Washington, 1996.
District Court for the Eastern District of Pennsylvania, ACLU v. Reno, June 11, 1996.
FC (The Unabomber), "Industrial Society and its Future," 1995.
Littman, Sol, "Some Thoughts on the Regulation of Cyberspace," Simon Wiesenthal Center (Canada), Toronto, 1995.
Newspaper Association of America, "Facts about Newspapers," 1996.
Ontario Court of Justice (General Division), St. Catharines, Ontario, R. v. Bernardo, 5 July 1993.
Sansom, Gareth, "Illegal and Offensive Content on the Information Highway," Information Highway Advisory Council (Canada), Ottawa, 1995.
Simon Wiesenthal Center (Canada), *The Need for Regulation on the Information Highway*, Toronto, 1995.
Subcommandante Marcos, "Neoliberalism: The Catastrophic Political Management of Catastrophe," July 17, 1995.
Southam Newspapers, Freelance Agreement, March, 1996.
Stein, George J., "Information War–Cyberwar–Netwar," New Dimensions International, 1996.
Swett, Charles, "Strategic Assessment: The Internet," Office of the Assistant Secretary of Defense for Special Operations and Low-Intensity Conflict (U.S.), Washington, 1995.
Tanaka, Tatsuo, "Possible Economic Consequences of Digital Cash," Centre for Global Communications, International University of Japan, Tokyo, 1996.
Yates, Larry Lamar, "The Internet: What it Can and Can't Do for Activists," presentation to Inet 96, Montreal, June, 1996.

Other Sources

Internet Documents

Anti-Defamation League Report, "Hate Group Recruitment on the Internet," 1995.
Beam, Louis, "Leaderless Resistance," *The Seditionist*, No. 12, February, 1992.
Internet FAQ for alt.fan.karla-homolka, August, 1993.
Macdonald, Andrew (William Pierce), *The Turner Diaries*.

McVay, Kenneth N., "Holocaust FAQ: Willis Carto and the Institute for Historical Review."

Smith, Winston (Harold Covington), "The Importance of Holocaust Revisionism," *NSNet Bulletin*, No. 5, July, 1996.

Interviews

Axtel, Tom, Taqramiut Nipingat Inc., June, 1996

Belsey, Bill, Leo Ussak School, March, 1996

Blumenthal, Robert, Clearnet Inc., October, 1995

Calisi, Chris, Symantec Corp., July, 1996

Cerf, Vinton, June, 1996

Choudhary, Aamer, June, 1996

Cleaver, Harry, University of Texas, November, 1996

Clegg, Frank, Microsoft Canada, April, 1995

Crossfield, Tina, March, 1994

Di Poce, Susan, Lotus Development Canada, October, 1995

Fockler, Ken, Canadian Association of Internet Providers, November, 1996

Fournier, Bill, Evans Research, October, 1995

Gagne, Louise, April, 1995

Gibson, William, August, 1993.

Gluck, Julia, Johnson and Smith International, October, 1995

Greenberg, Alan, McGill University, March, 1994, June, 1996

Greisen, Frode, The Internet Society, June, 1996

Hai, Ernie, National Computer Board, June, 1996

Handa, Sunny, McGill University, April, 1996

Hinckley, Peter, Systems Engineering Society, June, 1996

Johnston, David, McGill University, April, 1996

Jones, David, Electronic Frontier Canada, April, 1995

Katainak, Willy, Salluit, June, 1996

Kawchuk, Ron, Canadian Association of Internet Providers, November, 1996

Kleim, Milton J., September, 1994, May, 1996, October, 1996

Koenig, Tania, Aboriginal Youth Network, March, 1996.

Lake, Rici, Yellowknife, NWT, March, 1996

LeBlanc, Pascale, Promeve Inc., October, 1995

Littman, Sol, The Simon Wiesenthal Center, April, 1995

Livermore, Sid, IBM Canada, October, 1995

Macdonnell, Rod, The *Gazette*, September, 1993

Masse, David, September, 1996

McVay, Ken, The Nizkor Project, May, 1996

Morrissette, J.S., March, 1994

Nickerson, Ken, MSN Canada, November, 1996

Osmond, Gary, Royal Canadian Mounted Police, April, 1995

Parsonage, Keith, Industry Canada, September, 1996

Parsons, Neil, September, 1993
Paulson, Justin, Swarthmore, PA, November, 1996
Peloquin, Kathy, Taqramiut Nipingat Inc., March, 1996, June, 1996
Raven, Greg, Institute for Historical Review, May, 1996
Sadowski, George, The Internet Society, June, 1996
Saviadjuk, Paul, Nunavik Net, Salluit, 1996
Sikuku, Crispin, ThornTree Communications, Nairobi, June, 1996
Smith, Jason, Northern Hammerskins, 1994
Stupich, Tim, Industry Canada, March, 1996
Thompson, Murray, *The Standard*, September, 1993
Vallee, Vince, March, 1994
Ward, Karen, Concordia University, January, 1995
Weiss, Stanley, Novell Canada, April, 1995
Weitzman, Mark, The Simon Wiesenthal Center, May, 1996